ALL POOR TOGETHER

John Hollaway

Capricorn Books

Published by Capricorn Books
Johannesburg, Republic of South Africa
Copyright © John Hollaway, 2000

*John Hollaway asserts the moral right
to be identified as the author of this work.*

ISBN: 0-7974-2149-1

Originally typeset by Bruce Brine, Harare, Zimbabwe
This edition typeset by Margo Bedingfield
Printed and bound by: Colorpress
Cover design by: Di Deudney
Cover Photo: Dr Trevor Jones

Capricorn Books also published
The Burning of the Bankers
by John Hollaway

All rights reserved. No part of this publication may be reproduced ,storedin a retrieval system, or transmitted, in any form or by any means, electronic, mechanical, photocopying, recording or otherwise, without the prior permission of the author and publisher.

This book is sold subject to the condition that it shall not by way of trade or otherwise, be lent, resold, hired out or otherwise circulated without the author's and publisher's prior consent in any form of binding or cover other than that in which it is published and without a similar condition including this condition being imposed on the subsequent purchaser.

CONTENTS

Preface		iv
Acknowledgements		vii
Chapter One	FREE FALL	1
Chapter Two	A VERY GOOD BOOK	21
Chapter Three	THE STUDIOUS SHEPHERDS	44
Chapter Four	'LOOK ON IT AS A CHALLENGE'	59
Chapter Five	THE SONG OF A RED-EYED DOVE	75
Chapter Six	THE HEALING ART	89
Chapter Seven	A DISTURBANCE IN THE ROCKS	100
Chapter Eight	THE ROTTEN REEF	119
Chapter Nine	RUINED BY DRILLING	123
Chapter Ten	GOLD AND GOD	143
Chapter Eleven	SUITS AND SACKCLOTH	163
Chapter Twelve	RECKLESS ASSERTIONS	188
Chapter Thirteen	SMALL, SMALLER, ARTISANAL	205
Chapter Fourteen	IT'S GOOD TO TALK	225
Chapter Fifteen	WHAT A BUSINESS!	235
Chapter Sixteen	THE BARBARIC RELIC	243
Chapter Seventeen	SHEER CONJECTURE	258
Chapter Eighteen	SOMETHING IN THE AIR	280
Chapter Nineteen	NEVER ENOUGH	293
Chapter Twenty	IDA WHO?	308
Chapter Twenty-one	A THOUGHT AT SUNDOWN	324
References		348

Acknowledgements

An argument cloaked as an autobiographical fragment of this sort owes a debt of gratitude to hundreds of people who, one way or another, have contributed to its making. However, it is important to thank specifically those who have been kind enough to hone (or blunt) my ideas from their own knowledge and experience. These include Edward Nyamakeye, Joshua Nyoni and John Robertson. However, all the mistakes in this volume are mine.

Once we were rich and poor;
now we are all poor together.
[Applause]

— Julius Nyerere addressing a conference of Chama ChaMapinduzi, his ruling (and sole) political party, as reported in the Daily News of Tanzania, October 1982.

When people these days say
'our customs', 'our culture', you are
intended to take it seriously, even when they
are seeing these phrases like unsafe life-rafts in
a stormy sea.

— Doris Lessing, African Laughter, 1993.

Preface

> *The first task is to see the way in which our attitudes are rooted in the poverty, inequality and economic peril of the past.*
> —J. K. Galbraith, *The Affluent Society* 1958.

This is the story of a personal search for the answer as to why aid money — several hundred billion dollars of it—has made no impact on Africa's poverty and, infinitely more important, what would.

J. K. Galbraith was writing for the North America of the late 1950s, at a time when there was still some disbelief at the wealth that was washing over the inhabitants, and the fear of another Depression was not, in his words, a little thing. Galbraith's concern was this: 'to have failed to solve the problem of producing goods would have been to continue man in his oldest and most grievous misfortune. But to fail to see that we have solved it, and to fail to proceed thence to the next task would be fully as tragic.'

Galbraith was forty-nine at the time; from my sixtyish view of matters he was a bit young to be so Churchillian in style, and some of his conclusions are seriously dated. (He lightened up a lot afterwards; read *Money, Whence it Came and Where it Went.*) But as far as Africa is concerned he had it dead right in his fear that the lessons of Western economic history might be ignored. For Africa is a continent where past attitudes are driving millions of people into deepening poverty and present-day thinking on aid is equally inappropriate.

Its geology, or more specifically its quiet plate tectonics, has ensured that the soils of Africa are leached and infertile, while its status as the home of mankind has ensured that it is riddled with mankind's diseases. Without fertilizers it is necessary to move on every year or so. Coupled with a fearsome child mortality, in this environment the number of women, children and livestock are the assets by which a man is judged. This has led to communal landholding, polygamy, the bride price in livestock and, given a modicum of infant disease control, a population explosion. It also results in a deep, near-exclusive loyalty to the extended family that has been fatal to the development of properly functioning nation states in Africa.

All poor together

The lack of understanding of the African's adjustment to his or her environment—and its disastrous incongruity in a world of vigorous capitalism—has led to the greatest waste of taxpayer's money in peacetime history. During the fifty years from 1950 to 2000, about three hundred billion dollars was spent on aid—'development assistance' is the more dignified expression used by the 'donor agencies'—to the continent. During the latter part of this period, at a time when the flow of aid was at its height, the average African got no richer, and almost certainly has become poorer. Poorer by how much depends on whose figures you take, but overall probably of the order of five percent.

Short of building a bonfire of three billion $100 notes,* the money would have been far better used just by handing it out to the Africans, at a rate of around thirty dollars a year, for the best part of half a century. In very broad terms this would have given everybody in Africa an annual income supplement of between five and ten per cent; at the very least it would have stopped the inhabitants from becoming poorer.

During the same period, Japan and much of the rest of South-East Asia started on their march to success. The contrast was so marked that there was unconcealed glee amongst African leaders when, after decades of being nagged at by the International Monetary Fund and the like to emulate these paragons, South-East Asia underwent a massive financial crisis in 1997.

Yet at its most fundamental the Asian collapse was caused by success; after decades of unprecedented investment-fuelled growth there were not enough good projects left to go round. The money kept flooding in, so it went to dud schemes instead. When the inevitable losses occurred, the money took fright, regardless of the quality of its investments. Some primitive economic attitudes were in place in some of the countries, Japanese protectionism for example, but overall this was a financial crisis, not an economic one. Once the painful adjustments between reality and perception were made, all those Asian 'cultural' assets of hard work, thrift and commercial networking came into their own again.

Preface

At the beginning of this period, during which Asia grew and Africa shrank, I was at school, acting as an occasional spare pair of watchful eyes during the fortnightly gold 'clean-up' at my father's tiny mine in the then Southern Rhodesia. By the end of it I had become a too-much-travelled mining consultant. In most of the world mining is not seen as a serious business; too many little bags of purported gold dust have been pushed across too many Hollywood bar counters for it to be otherwise. But in Africa it is one of the few resources that is marketable. So with my modest skills.

Because I worked on my own, because I came from a background of small-scale mining (people like my father were known as smallworkers) and because Africa is a continent where something like fifty tonnes a year of gold comes from the activities of unsophisticated local gold miners, I became an involuntary, accidental even, specialist in such matters. An aid worker.

In this guise (which even now seems improbable) I found, as have many others, before and since, that aid was not effective in Africa. I was not, it seemed, part of the answer. Could it be that I was part of the problem?

Worse, the economists, formerly as one in their demands for liberal, open economies and fiscal discipline in Africa, were now at furious odds over what to do next. The only area that everybody agreed on was that aid does not work in a 'bad policy environment'. Like many of the phrases used in the development assistance business, this is the surface expression of a multi-layer code. The next stratum down attaches it to the writer's speciality—for the economists it means 'weak economic management', for technicians like myself it means 'the poor allocation of limited resources'. But below this there is consensus again; that it is due to fragile property rights, high levels of corruption and egregious taxation arrangements.

So now, with only one point of accord to build on, development assistance (which lives by buzzwords) has focused on 'governance' and 'civil society' as areas where special attention must be paid, so that aid will be effective. It can be predicted that shortly

All poor together

dedicated institutions will be created to inculcate the virtues they represent (just as they were for 'capacity building' and 'development economics').

The new focus is as doomed to failure as were the previous ones. The culture that has enabled Africans to survive for hundreds of thousands of years in their harsh environment will continue to be at odds with the fine visions of honest bureaucrats, equitable taxes and secure land tenure.

The way forward is not to try and change the way in which the system malfunctions from outside and from above. It is to start below, to construct a new African society that has an interest in reforming its own political and social arrangements for the better. This society would have the courage to leave behind the family-based, semi-nomadic, polygamous cultural survival mechanism that has become the instrument of its impoverishment. This society is already forming in a small way; my own African clients are members of it. Unusually for Africa they are invariably property owners.

The process? The French, naturally, have a word for it, *embourgeoisement*. It may have echoes of class warfare, but I have seen what damage a philosophy like that of Julius Nyerere's downward levelling has wreaked on the people of Africa. As Galbraith observed, our attitudes are rooted in the challenges of the past; we have trouble recognizing the elements of success when it is all round us. If we in the West have not solved the challenge of becoming 'all rich together', at least we have left behind us the major hazards of poverty, such as water-borne diseases and nutritional deficiencies. Why not Africa?

---

* *From page vi*
  A footnote in a preface may be ominous, but this is irresistible. Three billion $100 bills would weigh about five thousand tonnes, and their combustion would generate as much heat as the annual yield of about 10 square kilometres (4 square miles) of woodlots. Of course, if it was only dollar bills that one was dealing with ...

Chapter One

FREE FALL

> *... I was still mentally and nervously organized for war ... I retained the technique of endurance: a brutal persistence in seeing things through, somehow, anyhow, without finesse, satisfied with the main points of the situation.*
> —Robert Graves, *Goodbye To All That*, 1929.

We were dusty, unshaven and very tired, and stank of sweat and gun oil. The sharp tang of burnt nitrocellulose hung in the air. Seventy or so strong, we were sprawled against lichen-covered granite boulders under the thin shade given by the young foliage of the msasa trees. It is perhaps necessary by now to say that all of us in that group, at that stage of the war, were white.

All around the ripe seed pods on the trees were snapping and falling in the steady heat of the late afternoon, and a distant red-eyed dove purred, 'There's no gold—I told you so ... there's no gold—I told you so ...' Sometimes blackened leaves spiralled gently down amongst us from a hot grey sky. This dry season the bush fires were particularly bad, many lit by the terrs it was said.

For three days we had been training in this stretch of Rhodesian bush, attacking and being attacked, skirmishing, ambushing and, of course, waiting. Most of us had been through this pre-deployment refresher many times—this would have been, I think, about my twentieth call-up—and by now, August 1978, we had all experienced four years of the gnawing worry of living in a country—our country—suffering a guerilla war.

We were fighting terrorists, terrs. Recently we had been told we were supposed to call them CTs, Charlie Tangos, communist terrorists, presumably because it was a term associated with the successful counter-insurgency of the 1950s in Malaya. In Rhodesia it was a title that was only partly, only occasionally, true. To our enemies we were racist settlers, but that was not necessarily true either. This conflict was a turf war, the most common sort, but sustained by these handy labels of race and ideology. If these terrs,

All poor together

our enemies, were to have an abbreviation, then FGs would have been best, as away from polite company most of us referred to them as fucking gooks.

It was the final event of the training, an intelligence briefing. An army captain, intimidating in properly faded yet immaculate camouflage, stepped before us. Although we also wore camouflage, we were police, not army, in my case because eighteen years before my feet were considered too flat to march on by an army doctor. When serious fighting seemed imminent I decided it would be wise to join the Police Reserve, an outfit up to then known mainly for the beer busts at its annual training and the Christmas parties for the children of its members, in case the army reconsidered its decision. We watched in exhausted silence while the captain set up an easel on the sandy ground and unrolled a series of maps down it.

Much of the cant we used so easily then has left me, but his talk went something like this:

'The JOC where you chaps are going has fifteen known groups of terrs.' [He didn't say CTs either.] 'Total active about two hundred ZANLA, with AKs, SKSs and probably four RPG twos between them. There are also a few ZIPRA cadres, and we have reason to believe that they have at least one SAM-seven with them. However, Op Repulse has effective ground cover operating and as a result we have good int showing that they are in these locstats …'

After a lot more like this he got on to the fighting itself.

'Since May we have had seven successful contacts in the JOC, five killed, fourteen or so wounded, captured ten. RLI and PATU did it, mainly.'

One of us, a big unabashed farmer, had the temerity to ask what our losses were.

'Five, two from ADs. Eighteen wounded, about. Mainly from mines. Happy?'

'I heard there were twenty or so. Deaths.'

'Negative. I'm not counting road accidents.' The captain grinned. We laughed obediently. But accidental shootings and vehicle crashes caused many losses. Nor was he counting civilians,

Free fall

black and white. Anyway, in the place I had been going to, the Maranda Tribal Trust Land, we had manifestly been losing; three of ours to one of theirs in the last couple of months.

'So,' he went on, 'we think we are getting on top at last. Give it a year or so and we can start to wind down.'

We wanted to believe him, but we were neither young nor novices at the business, and there was a rumble of disbelief.

'OK ... let's have a look at the big picture. This war is not just about you people chasing a few gooks. That could go on for ever if the fucking politicians choose it ... it's outside Rhodesia that we'll win or lose.' The map of the area where we were to be posted, and one or two of us to die (at this distance I have even forgotten the details of our modest casualties), was flipped over the back of the easel, revealing that of sub-Saharan Africa.

He pointed to Lusaka.

'The terr unity thing is finished. There is no longer a Patriotic Front; it went the way of Frolizi. Nkomo and ZIPRA rule the roost in Zambia with Russian support. Do their training beyond Solwezi, up here. Mugabe and his ZANLA in Maputo. Chinese assistance. Their training camps are in Tanzania as well as Mozambique, here, here, here. So now two rival armies. We dug up six ZANLA terrs from the Op Tangent area two weeks ago. All shot in the back of the head with the same nine-mil Tokarev. Kaunda in Zambia, Machel in Mozambique, Nyerere in Tanzania, they are totally pissed off. There was a meeting in Dar-es-Salaam recently, where we hear they were seriously talking of withdrawing their support for the terrs.'

I don't know what anybody else thought, but this last seemed unlikely.

'Why? Why, because they can't afford it. Their economies are in Free fall. These people couldn't run a bar in a brewery. They've had enough of carrying the gooks on top of everything else.'

The moment has stayed in memory, for he was talking in front of a big munondo tree with a speckled crown of large seedpods. I stared absently up at it, the long-unused words Dar-es-Salaam recalling a girl, a woman actually—now quite an old woman, I realized in surprise. Everybody was saying what a shambles these

All poor together

countries to the north of us were. I wasn't sure, perhaps because, unlike anybody else in that group, I had actually been there. Fifteen years ago admittedly, but I remembered great optimism and an impatience for independence, which was only months away then. The country had been full of—what were they called?—development assistance experts, enthusiasts full of theories and schemes.

Suddenly one of the pods I was watching jumped apart with a crack like—well, like a Tokarev pistol. The two halves spun down, leaving behind a brief, flickering trail of seeds in free fall.

· · · ·

Luck. But for it I would not be writing this and you would not be reading it. Not one piece but four.

The first. In early 1982 I was introduced to Bo Eriksson, a Swedish geologist. I had left my employer six months before to become a mining consultant, but I was retained by them for advice on coal. They had discovered a very large deposit shortly before the guerilla war had hotted up, and now, two years into peace and Zimbabwe's independence, wanted to develop it.

Bo, whom I was to come to know as the very best sort of Swede, worked for a huge state-owned iron ore company, which was seeking coking coal at that time. So he met me in my ex-employer's head office, a tall building in Harare—still called Salisbury then—to hear about the Sengwa coalfield. Unfortunately, the coal seam there is not only remote and undeveloped, but it is non-coking, so our meeting didn't last long. I got ready to drive back to our home, 130 kilometres to the south.

'Were you in the war here?' Bo asked as I rolled up the maps.

'Off and on.'

'How was it?'

It was a question that could be dealt with only by a word or a book. 'It was a laugh a minute,' I said.

'Oh ... I was working on an aid project near the Chimoio camp when you attacked it. A message came telling us to get out a couple of days before.'

Free fall

The Chimoio raid had been an enormously successful Rhodesian ruse; hundreds of would-be terrs had been gunned down as they stood on parade to welcome what they thought were reinforcements. Bo Eriksson had probably been in much greater danger than I was. I have often wondered who warned them.

'Are you only a coal consultant?' he asked.

'My main work is in gold. We've a lot of small gold operations here. I do assays, tests, sort out problems on the mines. Tell them what to look for underground.'

'You do? You know we have a gold mine in Tanzania? Up country, rather a long way. You fly from Dar-es-Salaam and then you have to cross part of Lake Victoria on a ferry to get to it. A day's trip. Very bad roads. Our people have some trouble with malaria.'

It sounded not much fun. I gave him my business card.

A couple of weeks later a long telex came. Could I go and look at Bo Eriksson's mine? It was called the Buckreef. They were having gold recovery problems in the mill. Perhaps not enough leaching time. I was to let them know what my fee would be. I was to liaise with Hekki Sorensson in Dar-es-Salaam. He was with the State Mining Corporation there, which was called Stamico. At your very earliest convenience, please.

Buckreef was an unusual name. Many, perhaps most, gold mines in Zimbabwe are called after wives or girlfriends, such as, just to take just one letter of the alphabet of mines in Zimbabwe that I have been on, the Ella, the Emma, the Ethel and the Ettie. Others are given grander names, sometimes after royalty (the King Edward, the Princess Royal). Some are hopes (the Eldorado, the Midas) or fears (the Last Chance, the Goodbye), some combinations of these (Maggie's Luck, Royal Mint). Many names, though, seemed to be nothing more than a private joke (Step Lively, Nubobs, Thistle Etna). But a buck reef was a miner's name for a barren white quartz vein, a mine without any gold in it.

There was no question of my not going, distance, malaria, bad roads and all. I had cashed in my pension to start this business, only to find that gold miners who needed help usually did so because they had already spent all their money on the wrong things.

All poor together

Other people's financial problems rank well below holiday snaps in audience appeal but, briefly, if I didn't get some serious money in very soon, I had about two months left to find a job of the normal, day after bloody day, sort. Whatever else might happen in Tanzania, I should be able to count on getting my fees paid. It was a major stroke of luck; perhaps it was fortunate I did not know that I was going to need three more of them in this job as well.

Hekki Sorensson replied to my telex with a suggested flight that would take me to Dar-es-Salaam and onwards the next day to Mwanza, on the shores of Lake Victoria. He also asked for my phone number. I gave it to him with the caution that it was difficult enough to get through from Salisbury, let alone Dar-es-Salaam. It was a party line, and when it worked it seemed to be monopolized by farmers' wives swapping cake recipes, or their offspring entangled in paroxysms of giggles. Most of my clients, despite being on their uppers, didn't worry about even trying to phone, but came bouncing up to my office-cum-laboratory in their battered pick-ups.

Nevertheless, about two days before I was due to go, our four-longs-and-a-short called me to the telephone and, through a rush of static and a babble of foreign voices, a distant, clipped, female one told me to hold on. If asked to say where I was connected to, I would have guessed a busy Danish fish-and-chip-shop, but then somebody announced that this was Hekki here. He was calling from the Finnish Embassy in Dar, he said, because their phones worked. He was glad he had finally managed to contact me, because he had an important request from Pat Carter.

'Pat Carter?'

'He's the General Manager of the State Mining Corporation. It's about Buckreef. The plant has a problem because they have no nitric acid.'

I considered this briefly. Nitric acid is used for a few things on a gold mine, for assaying, where it removes silver before weighing the gold prill, and for cleaning the bullion bars after smelting and pouring them. Important tasks, but not critical, normally.

'Can you bring acid with you? Mr Carter thinks you probably

Free fall

won't be able to do anything unless they have it.'

'What do they need it for? How much do they need …?'

But a fresh big basket of raw chips had been plunged into the hot fat, and then, with a faint pop, everything went quiet.

Now for the second piece of luck. If this was a primer on becoming a mining consultant, which it isn't, this is the part where I would emphasize the great importance of having a good travel agent. Mine (who will be anonymous as I still need her undivided attention) had handled my flights even before I launched into this consulting business. When I asked her to book me to Dar-es-Salaam, she began to issue travel advisories.

'Air Tanzania is one of Those Airlines. You never can be sure what is going to happen. Sometimes they just don't come. Make sure you have jabs for yellow fever and cholera, otherwise they will try and do it to you at the airport. Somebody said you should take your own toilet paper. Don't drink the water.'

But her really valuable service was to manage to contact me just the evening before I left. Over a line which clicked agitatedly as others on our circuit registered their impatience to get on, she said that it looked as if the flight was going to leave a couple of hours early. Maybe it wouldn't, but best to be on the safe side.

All unknowingly she made—or at least saved—my career with that phone call. For the flight was early, departing at 10 a.m. and not midday as scheduled, and thanks to her I was on board. Air Tanzania Corporation (ATC, Air Total Confusion, the Tanzanians said) was not big on timetables. Its schedules were sparingly distributed on a single, blotchy roneo'd sheet that had to be modified and replaced almost daily. In Tanzania it was visible to aspiring passengers only under the glass covering the booking office desks, a challenge in upside-down reading and a source of twisted necks.

This nitric acid business. I discreetly broached it with my travel agency lady, and after she had done some falling about (forget it, they will never let you take that up, she said) she hunted out the requirements. Aeroplanes are made of aluminium, as are the containers of bulk concentrated nitric acid, but that was not how IATA saw it. Effectively it was as forbidden as a loaded gun.

All poor together

But for all I knew, this Mr Pat Carter person might send me back if I hadn't got the stuff, and if he did that then I might as well start applying for salaried employment right now. So I bought five litres of analytically pure acid in a sealed clear plastic container and removed the label. After eight years in an environment where the ends justified the means, I was going to show these Tanzanians that I was a can-do sort of fellow.

There was no x-ray machine at Salisbury in those days; passengers were called in one by one into little cubicles and were hand-searched. The elderly black security guard who checked me—probably an ex-BSAP policeman himself who had decided that the time had come to get into less politically sensitive employment—could not help but find the plastic bottle.

'What's this?'

'It's a lixivant.'

'What?'

'A lixivant; a chemical used in mining. It's pretty harmless—look.' I cracked open the seal and unscrewed the top. He peered in and sniffed. Pure nitric acid has no smell to speak of.

'Are you sure it's safe?'

'Of course.' I put a finger into the cool, dense liquid, pulled it out and showed him.

'Okay?'

He nodded. I resealed and repacked the bottle and carried my hand luggage through the curtain into the departure lounge, moving smartly towards the toilets. The finger was stinging fiercely when I got to the hand basin, but after a few minutes sluicing I was left only with a deeply yellowed digit and a sharp, persistent pain under the nail.

It was a silly thing to do, criminal even. A security man less impressed by words he didn't understand (lixivant means a leaching agent) would have asked me to explain exactly what I meant, with ramifying consequences. But luck—the third piece—saw to it that he didn't.

When I got off the smelly, battered 737 with the giraffe on its tail and into Dar's humid heat, I was met by Hekki Sorensson, a youngish Finn who was carrying a sign with two names on it.

HOLLAWAY VERGYMPF.
'Where is Mr Vergympf?' he asked.
I had never heard of him.
'He's another consultant,' explained Hekki. 'He should be on this flight; he was coming from Namibia to go to Buckreef with you. Stamico arranged it.'

We hung around on the periphery of the crowd that was edging its way past the health and immigration checks, but nobody else came up to us. The queues dwindled and vanished, and I took my bag to customs. The official there did a brisk rummage through it and triumphantly pulled out the bottle of acid. Hekki caught my eye as I started my 'just a lixivant' routine again. Luckily this time I did not have to prove that it was harmless, or even dutiable, and after a further cursory glance at the other contents (mosquito net, sleeping bag, gold pan, duty-free whisky) the bag got a chalked cross and we walked out to where a driver was waiting with an elderly Land Rover. The word Stamico was roughly stencilled on the front doors.

'Was that the nitric acid?' he asked.
'Yes.' I showed him my yellow forefinger. 'What do they want it for?'
'The laboratory, I think. Pat Carter knows about it.'

Dar-es-Salaam seemed much bigger and messier than I remembered from the last time. That was over twenty years earlier, when Tanganyika was still the name of the country, and I was a callow twenty-something, bumming about Africa. Great piles of rubbish had been built up on the pavements. The battered buildings and public places of the city seemed to have been under a long siege, just as I had seen in Maputo and Beira. However, Mozambique had been directly involved in our war, unlike Tanzania.

'Pat Carter wants to see you before you fly up to Mwanza tomorrow,' said Hekki. He also had particularly wanted to see Mr Vergympf, who was from de Beers, the diamond company, in Johannesburg. Despite sanctions against the South African racists, some sort of discreet connection was still maintained by Tanzania with them, because of a famous diamond mine called Williamson's,

All poor together

after the discoverer. Until Buckreef came along this was Tanzania's only serious mine, in a practical sense (if not in an economic sense). De Beers had owned it until they were reduced to minority shareholders at about the time of Tanzania's independence. Now their interest was discreetly hidden behind a Canadian company.

Bo Eriksson's firm was providing technical support to Buckreef through a Swedish aid programme. However, the mine was actually owned by a government company, the State Mining Corporation. Any mining activity in the country was supposed to be under the control of Stamico. I remembered that there had been a score or so of gold mines in Tanganyika when I was last there; now it seemed there was only one, Buckreef, and evidently it was in trouble. Mr Vergympf had been summoned, ostensibly from Namibia, by Stamico. I had been recruited from Zimbabwe by Bo Eriksson. In the event I never met Mr Vergympf, so I don't know how good he was at his job, but I had the good fortune to have the better travel agent.

Hekki, who was working for Stamico on Buckreef (through a Finnish aid programme) told me about this competing consultant over a Safari pilsener (Czech aid programme) in the down-at-heel lounge of the Kilimanjaro Hotel (Israeli aid programme until the seven-day war). Before Hekki arrived at Stamico there had been a whole series of earlier Buckreef aid programmes. They included, if I remember correctly, the Swedes, the British and the Canadians.

In spite of, or perhaps because of, this sequence of experts, Buckreef, a modest mine by any standard, had taken ten years to build and had cost, it was said, ten million dollars. More, although it had been completed six months ago, it had not managed to get properly started, as yet. Most new mines in the bush take longer to build than their Gantt charts predict, and most have things go wrong on them on start-up, sometimes big things, but this seemed to be breaking new ground.

So that was why I was here, and why Mr Vergympf should have been. We had been recruited to deal with a matter of national importance. Tanzania's sole gold mine was in trouble. National prestige was at stake. President Nyerere himself, the Mwalimu, was starting to ask questions.

Free fall

At least, that was how Pat Carter described it when he eventually arrived. He was a man of mixed race; a English father and a Tanzanian mother. Years later one of the Finns, a blunt member of a candid nation, summed up their opinion of him. 'Oh, that one. All he liked was beer and big tits.'

It was growing dusk, and Pat Carter was concerned that I get an early night.

'Just a quick drink ... the flight to Mwanza leaves at seven tomorrow morning. You have to be at the airport at five.' Air Tanzania knew the dilatory habits of its passengers.

'I've brought the acid.'

'The acid?'

'Hekki tells me you wanted nitric acid.'

'Oh ... yes. For the lab. Can you fly up with it tomorrow?'

I resisted this. 'What do they need it for?'

'The laboratory furnaces are not working. The chemist there says he needs the acid instead.'

This left me none the wiser.

'What is he going to do with it?'

'For the assays. I don't know the details. They need it to dissolve up the samples, something like that.'

'Look, I'd rather leave the acid for Stamico to get up there. I could get into a lot of trouble if I was found with acid on the plane. I don't want to take the chance again.'

'But what about the assays?'

'Let me see what the matter is first. There is probably another way around the problem.' Under the nail my finger still stung intermittently, reinforcing my obstinacy, and he had to be satisfied with that. Anyway the beer had arrived.

I had noticed that there were no labels on the bottles of Safari. It turned out that this was because there was no foreign exchange to buy them. The bare metal Crown Cork caps were levered only partially off, to prevent flies entering. Before drinking people lifted their bottles up to the light to check that there were no foreign bodies. Still, although the glasses were cracked and greasy, the Czechs had done pretty well on the important part.

'Well,' Pat Carter said, cheering up after a long swig, 'what

All poor together

about a quick bite to eat? We can't be late, because of your flight tomorrow.'

So we ate in the restaurant at the top of the Kili. The view out over the sparkling lights of the shipping in the docks was excellent, but the food was poor; in a city where fresh fish is readily available, mine was several days old. Pat Carter also insisted I try Dodoma Red (Italian aid programme), a truly awful wine that must have undergone a major change from the time when it was tenderly fermented for their own table by upcountry Italian missionaries.

It emerged that I was to fly up with Hekki's wife. She had never been to Mwanza, the nearest centre to Buckreef, although her husband had practically lived there on occasions. A driver would collect her at 4.30 a.m. and then come round for me at about 4.45. Hence the need for an early night.

The service was slow, but eventually we finished up with Konyagi, a local raw cane spirit related to cognac only by its alcohol content, but not bad with Double Cola, the local substitute for the Real Thing.

During the evening, as the Safari beer went down, Pat Carter had been livening up.

'Hekki, let's take Mr Hollaway to Margot's. Just for a short time, as he has to start early.'

Twenty-odd years before I had known Margot's as a smart restaurant whose *maître de'* was the first ostensibly gay black I had ever encountered. This was not the same Margot's at all; it was a grubby night-club in the dock area where girls were, well, not just available, but abundant. We drove in Hekki's Land Rover through the dark, potholed streets and parked outside a run-down multi-story block. A shadowed figure came up at once and Hekki negotiated for him to guard the car while we were inside.

I had first seen this arrangement in Addis Ababa in the early 1960s; it had evidently spread south to Dar-es-Salaam by this time. A few years later in Harare the level of car theft had also risen to the point where casual parking meant paying for an unofficial guard as well. Now there is probably no city in Africa where you can park your car without a street kid or, more likely, a street youth

Free fall

approaching you to offer the service.

I was overdressed for this sort of thing. 'Leave your jacket. Take off your tie,' commanded Pat Carter before we got out. We pushed our way through a group of bouncers and whores and climbed up a couple of flights of grimy, narrow stairs. At the top we paid a largish entrance fee and entered a big, dark room that was full of noise and girls. The latter were dancing in ranks by themselves; there seemed to be no other male clients at the moment. The girls were of every colour except white.

Pat Carter made a view-halloo sort of noise, plunged into this shadowy throng and vanished. It was three months before I was to meet him again. Hekki bought me a whisky, then politely excused himself and left for the dance floor as well. Carter might have disappeared but at least I could see Hekki, bouncing away in the crowd. Emboldened, I swooped in as well. Just like a hop at the mine club, I reassured myself. Music a bit noisier though, more ethnic. Then I became aware that Hekki had vanished as well, and I retreated to a chair by the wall.

Here I sat for what seemed to be several hours, beleaguered by importuning women. My nervous, dogged resistance led them to insist that, rather than waste their time, whiskies had to be bought. All round. I don't know what they were given but I got small doses of a nasty version of it for very large numbers of shillings. I began to get really worried.

A digression. When Robert Mugabe took over the government of Zimbabwe he inherited a siege economy designed to resist the comprehensive trade sanctions imposed by the United Nations on Rhodesia. This meant that even the least decisions involving foreign exchange had to be approved by bureaucrats in the Reserve Bank. Conveniently, such an arrangement is just the thing that a socialist government such as his requires. As a result (and this explanation was so difficult for any outsider younger than about fifty to comprehend that it was usually not worth even trying) it took several weeks to approve the purchase of hard currency for a business trip, or rather longer than under Rhodesian sanctions. To get here (more can-do) I had cashed the family's annual holiday allowance, all of two hundred U.S. dollars. Now

All poor together

I was buying five-dollar tots of whisky for half the whores in Dar-es-Salaam, the wild-eyed centre of a small, excited crowd.

Suddenly Hekki broke into the circle with a dishevelled little black girl grinning on one arm. 'Let's go. You're off early tomorrow.'

'What about Mr Carter?'

He shrugged, and with profound relief I followed the couple down the cluttered stairs and out into a night that seemed remarkably cool and fresh. Hekki gave the car-watcher a few shillings and drove me back through empty streets to the Kili, where he and the girl left me. I had spent over half my family's holiday allowance for 1982.

* * * *

After hours of travelling, most of them in the dark, Sven the miner, who was driving, leaned forward and pointed as steadily as the bucking vehicle allowed.

'Dere's der mine.'

I bent down from the back seat and peered past him. A bright point of steady light shone in the distance out of the blackness, but then flickered out of sight behind the trees as we slid awkwardly down a sudden slope. It was the first electric light we had seen since leaving Mwanza five hours ago; apart from the occasional dim flicker of a cooking fire, the night had been like a coal hole beyond our headlights.

Not even another vehicle had shone past us on the road. It was a Sunday, and in 1982 the Middle East oil crisis—more correctly the oil price crisis, for there was plenty of the stuff about—was still making its own contribution to the crippling of the Tanzanian economy. Sunday travel was forbidden without a special licence, which we had to produce to the police at the three languid roadblocks we had encountered on the rutted road.

'After every trip you feel as if you haf been in a big boxing fight,' Claus the engineer had said feelingly, and he was right. The ferry ride across the arm of Lake Victoria at Mwanza was the only smooth part of the trip. The road we had been on before we turned off southwards to Buckreef was nominally one of the major arteries of Tanzania, running around the base of the Lake and then north up to the coffee fields of Bukoba and the Ugandan border. But it was in desperate need of maintenance.

To Tanzania's partial credit some of this was due to the heavy military traffic it had carried a few years before. Idi Amin, then the psychopathic dictator of Uganda, had staged a bloody comic-opera of an invasion of the Bukoba district, only to be repulsed, invaded in turn, and finally unseated by the Tanzanian army. I was far away, involved in another war at the time, but anyway the Tanzanians replaced him with Milton Obote (the legitimate ruler who had been supplanted by Idi Amin) who, in the event, turned out to be almost as murderous. Nonetheless there are too few cases in Africa of the right thing being done, and this effort deserves to be remembered.

To digress a bit further, the credit to Tanzania above is grudging because I found out later that over 20 years the country had received two billion dollars in aid for building roads. Some of that money must have gone into the occasional abandoned piece of earth-moving equipment that flickered past us on the verge, each with a tiny camp-fire tended by a lonely guardian; there was no sign of the rest of it.

Twenty minutes after that first hopeful gleam the lights of low houses, proper houses with corrugated iron roofs, appeared in the forest about us, and Sven turned off the track and drew up in front of one. The bulb over the front door was a silvery vortex of insect life.

We got out and stretched with relief. Around us the other houses, dimly lit or altogether dark, were silent. The air was fresh and cool—it was the end of the rainy season—and apart from the trilling of cicadas only a distant throbbing could be heard, which must be from the generators. Göran (whose name was pronounced Yuran) was the third Swede in our little group; he was in charge of the mill, and he spoke little and smiled not at all.

'It's pretty quiet,' I said to him. He shrugged. A mine is usually a noisy place, particularly at night when the silence of the bush is being interrupted by the trill of the shaft bells and the intermittent roar of fresh ore coming down the chutes, to be followed by a

All poor together

sudden champing as it enters the crushers.

'Everyone sleeping. The power it goes off in one hour at ten,' said Sven. We filed inside.

Their house—to distinguish it from the rest it was called, inevitably, the Swedish House—was a basic one. 'Concrete block under iron' would be a Zimbabwean estate agent's description. The floors were bare concrete. The main room had a table and a few sticks of locally made furniture; the table legs sat in tins filled with paraffin to stop the ants from climbing up to the food. On the door into the kitchen was a calendar, with each day marked off by a cross. Claus went over to it and brought it up to date; the Swedes had been away for three days over the week-end, staying with their families in Mwanza.

The cook had not known we were coming, so we supped on some of the bananas we had bought at the side of the road and drank unsweetened black coffee. Then Claus drove me a hundred yards to another, similar, building called the guest house, where I was to sleep. The water from the shower was cold, but there was a bed with sheets even, so my sleeping bag was unnecessary. I used my belt to sling my mosquito net as I did not trust the rusting gauze that covered the windows, and slid gratefully under the musty canopy and between the coarse sheets. At that moment the throb of the generator faltered and died, the bare bulb above flickered and dwindled out and pure silence swept in. For a second I started to wonder what adventures tomorrow would bring, but sleep arrived before the thought was completed.

· · · ·

The Arabs say that the soul cannot travel faster than a running camel, a piece of romantic nonsense probably put about by an early sufferer from jet lag. Nonetheless I awoke to a brief, disorientated panic. Shaded daylight filtered through the mosquito net, showing an unfamiliar, bare room, yet with a familiar dawn chorus of African bird song outside, tip-tolls and doves.

Then the generator started again, and the comfort of memory returned with it. After a few seconds the pale light bulb above came on with a yellow glow. I ducked out from under the net, had a cold wash and shave and unlocked the door into the cool, young day.

The squat, grimy house behind me was in the edge of a forest composed almost entirely of tall mfuti trees, their fronded leaves forming a high canopy. In front was a cleared area of several hectares in which Buckreef Mine had been built. A headgear towered above the mine buildings, the winder wheels stationary. Workshops and offices of corrugated iron bordered the operation, and I could see the conveyors and tanks of the plant, where my business would lie today. There was nobody about.

I followed last night's Land Rover wheel marks back along the muddy track to the Swedish House. A quiet breakfast was already in process; the mpishi—the cook—had appeared and was producing plates of scrambled egg to go with paw-paw and coffee.

'So the mine is shut down?'

Claus nodded, his mouth full. Sven said, 'I haf mined a little bit out, you know, about two thousand torns, but the stockpile it is nearly full now.'

'You will see,' said Göran, and the subject was closed.

We drove the two hundred metres to the mine entrance. A guard at the barrier peered into the vehicle suspiciously, and I had to get out and be signed on as a visitor. Around the primitive guard house children were playing. 'Hey hey!' they cried out in delight, the Swedish greeting. I searched my memory, and came out with 'Habare?', 'What news?', the Swahili greeting. They giggled and scattered.

Ranks of corrugated iron shacks for the workers and their families stretched away on one side of the entrance boom, down to a marshy stream. Amongst them people were stretching and stumbling about, with a first-thing-in-the-morning look.

The vehicle bounded along a rutted track through the mine to the offices, a substantial house that seemed to be of an earlier era. Outside it, raised on blocks, was a very, very old Land Rover, a flat-sided little Series One whose extreme age was discreetly revealed to me, when I had a chance to look at it closely, by the Whitworth thread of its bolts.

All poor together

This vehicle belonged to the Tanzanian mine manager, a comprehensively depressed mining engineer whom the Swedes called Mr Sucka, 'Mr Sigh'. He had been most unfairly treated; he had been in jail (oh heavens, a Tanzanian jail!) because of what sounded like some mild foolishness over a few gemstones, and on coming out had been given this appalling responsibility. Like almost everybody there, he craved only to be sent back to the *dolce vita* of Dar-es-Salaam.

He was away at that moment, in Mwanza dealing with some heavy paperwork for a big new generator. The silent Göran took me to the plant and introduced me to its Tanzanian manager, a stocky Machusa from the Livingstone Mountains in the far south. Machusas have a reputation as good, tough miners and I had met many on the Copperbelt and even down in Rhodesia, working underground on the Wankie coal mine. Mike Lutula had served a hard apprenticeship at Williamson's diamond mine before being given this job and, while he knew little about gold mines, he could make people work, the first of the mining virtues.

In fact there was almost nobody there who knew very much about gold mines. Of the three hundred Tanzanians on the Buckreef payroll, only a tiny handful were relics of the period, twenty years ago now, when such mines existed in the country. The Swedes had no gold background either; they were from the mighty iron ore mines of LKAB, far in the frozen north. It was hard to conceive of experience less relevant to a small gold operation in Africa.

It was my fourth piece of luck.

Göran and I walked down to the silent plant. To get gold out of rock, the plant used cyanide. However, to achieve this the rock has to be ground down in water to the consistency of mud, and then the thick grey stream that has been created is mixed with cyanide. To get big lumps of rock from underground, some thirty centimetres or more across, into this state is very laborious; most of the power from the generators was consumed by it.

Cyanide is, of course, a very dangerous chemical. I know of half a dozen successful and attempted murders and suicides that used a tiny amount (about a fifth of a gram or hundredth of an ounce) from stocks that were originally destined for dissolving

Free fall

gold. Indeed, the first (Tanzanian) Chief Engineer at Buckreef had died suddenly from, it was said, cyanide in his tea. A crime of passion, it was also said. From that point of view the four or five thousand tonnes of cyanide that a little gold mining country like Zimbabwe uses annually could kill not just the world's adulterers but its entire population.

At that time both Sweden and Tanzania were alike in that they had no gold mines. So at Buckreef, when all was ready after ten years of exertion, somebody asked—it may have been Hekki, for there were no gold mines in Finland either—'What do we do if somebody has an accident with cyanide?'

The practical answer is that somebody who accidentally swallows the very dilute cyanide solution used on gold mines can be revived by being fed common chemicals, either sodium thiosulphate—photographer's hypo—or a mixture of sodium carbonate and ferrous sulphate. Bottles of these antidotes sit in red-painted boxes on every gold plant, gathering dust, because accidental poisoning is almost unheard of. I once asked a doyen of the industry, one who had served almost all his working life as a regional mining engineer in the old Rhodesian Ministry of Mines, if he could remember such an incident. After a lot of thought he dragged a forty-year-old tragedy out of his memory. An African child—a picaninny, as he said—was sent from one small mine to another with a bowl of cyanide briquettes on his head. It was during the rainy season, and in a downpour the bowl flooded and the concentrated solution dripped down and into the child's mouth. It could not have been a nice death; cyanide is caustic.

That tragedy apart, a survey of ten years of accidental poisoning cases in Zimbabwe (1980–90), covering over 6,000 incidents, did not list cyanide.[1] Paraffin and pesticides were amongst the most common causes. Cyanide should be a threat; it exists in the cassava plant which, with maize, arrived on the continent from South America in the early sixteenth century and is now a staple food in much of Africa. Its root can be ground to give a flour, called manioc in its homeland (in Brazil, mandioca), and unless this is well leached in water or exposed to the sun to remove the

All poor together

cyanide present, it can be a killer. But again, the survey did not mention this amongst the food poisoning cases. Disappointingly for the environmentalists, cyanide turns out to be a hazard only in the hands of those wishing to make it so.

But at Buckreef the Swedish medical authorities were consulted and they recommended an oxygen canister and mask. Accordingly, the mine start-up was delayed for two months while the oxygen equipment was located and acquired. In fact oxygen is only handy if hydrogen cyanide gas is likely to be inhaled. It is not much use on mines, where the poison is found as a stable solution. Buckreef still had no antidote for somebody who had drunk the stuff.

Finally the mask and gas cylinder were delivered and, about five months before I arrived, the start-up commenced. The headgear wheels spun, the crushers began munching away, the grinding mills starting turning and the pulverized rock, in the form of a dilute grey mud, began to pour out of them and into the big mixers where the cyanide was added to dissolve the gold.

And that was as far as it got. The gold had still to be extracted out of the slurry, and the elaborate arrangements for achieving this did not work. Göran's experience on iron ore plants and Mike Lutula's on diamond recovery were of little use. For months they had worked frantically, in growing desperation, trying to make the process operate. Even the languid officials in the offices of Stamico and government were issuing sharp reprimands and, worse, unsolicited and urgent instructions as to what should be done.

This advice had failed as well; Buckreef had produced no gold, none at all. Göran's reticence was not just because he was naturally quiet; his was the silence of despair.

Chapter Two

A VERY GOOD BOOK

> The disaster was the product of the system rather than of the men themselves. The worst that can be said of them is that they were not very good engineers.
> —*Nevil Shute on the R101 airship catastrophe of 1930.*
> *Slide Rule, 1954.*

The technical term for the slurry of ground-up gold ore and weak cyanide solution is, memorably, pulp. Both Bo Eriksson in his telex and Pat Carter in his cups had said something about perhaps there not being enough time for the cyanide to finish its work on the gold in the pulp. They were baying after an old scent again, but only because all the others had proved fruitless. The six huge pulp mixers at Buckreef—agitators, they are properly called—gave nearly a week's leaching time. Twenty-four hours is normally more than enough, so if there was any leachable gold in the ore, it should be well and truly in solution by the time the pulp left the agitators. Göran and Steven had been fighting other battles, far more intractable than the pulp leaching time, for the past five months.

The system that had been adopted at Buckreef for separating the gold-bearing solution from the pulp was, at first sight, conventional. The pulp flowed into settling tanks called thickeners. The thick mud coming out at their base—the underflow—was pumped to rotating drum filters. These were covered with canvas, and a clear solution was sucked out of the pulp through this filter cloth, leaving behind a thick paste which was scraped off automatically. This gold-bearing cyanide solution (pregnant solution, it was called) joined that coming over the edge of the thickener, and the whole lot went through a clarifier to remove any remaining particles. After this, in theory, gold could easily be precipitated by putting another, more reactive, metal into solution, in this case zinc dust.

Steven escorted me around the plant. Göran stayed behind in the poky manager's office there, to deal with some administrative work he said. I am not a perceptive type; writing this now, years afterwards, I realize he probably could not face taking me, the parachuted-in expert, around the arena of his defeat. Göran, who for all his hard work was being accused by the Tanzanians of losing their gold, eventually left to join a Swedish iron ore company.

'At least there I am not likely to be called a thief,' he told me bitterly before he left. But his new job was in Liberia, which collapsed into bloody chaos some years later. I hope he got out safely.

For now we are still in 1982, and Steven and I have climbed to the platform on top of the huge agitators (quite amazingly big, for the size of the mine). We are fifteen metres up. Only the headgear is higher than we are, and our view is of the forest canopy stretching away to the far horizons about us, in a measureless succession of gentle dark green crests, motionless under a high African sky. Twenty years ago, as a young assayer, I had stood on a hill near Luanshya and marvelled at the interminable canopy of the *miombo* woodland that surrounded the Zambian Copperbelt. That was sixteen hundred kilometres to the south; Buckreef was another mining island set in the same forest, the same trees. Tree, really, for the dominant species is the fronded mfuti, which we whiteys call, clumsily and eurocentrically, but accurately, the Prince of Wales Feathers tree.

Steven pointed out the various features of the sprawling rusty plant beneath us, the setting of his and Göran's futile, five-month-long, campaign to produce gold. There was the pipe that had split and nearly blinded somebody, there was the thickener they had spent a week digging out because a wrench left in it had jammed the mechanism, there were the rotating drum filters whose cloth would never align properly for more than an hour at a time, there, most frustrating of all, were the fancy pressure filters which were supposed to clarify the pregnant solution, but which choked within minutes of every start-up.

These last were new to me, although I kept a modest silence about this. This static filter section was the final run-up to the precipitation of gold from the cyanide solution using zinc dust, and it looked like ... well, not at all like a gold plant clarification section, which is usually a couple of tanks with sand in their bottoms. These filters were a rank of sealed, domed vessels with massive bolts holding their tops on, embedded in a maze of pumps, pipes and valves. I had not the foggiest idea what went on inside them. If anything they resembled part of a modern brewery.

On the other hand the thickeners were conventional enough. They were big, squat circular tanks full of pulp, which was supposed to separate into mud and a clear overflow in them. They seemed to be very big for the size of the plant; perhaps twice the diameter I would have expected. Hello! I could see what looked like jellyfish drifting on the surface. Now here was something I did know about. We climbed down from the agitators and up again on to the walkway over the thickeners. Steven saw me looking at the amorphous pale blobs.

'The flocculant did not mix, it just made lumps.' Flocculants are needed to agglomerate fine clay particles into bigger ones, so they can settle and leave a clear solution to come over the lip of the thickener tank and into the channel placed there to carry it off to the Static Filters. Steven's flocculant had not done this; it had just formed big blobs of infernally sticky white jelly. The surface of the pulp in the thickener was soup-like from the clay particles in suspension.

The problems at Buckreef were boiling down to a failure to separate liquid from solid. The thickeners didn't work, the drum filters didn't work, and because of that the mysterious static filters apparently didn't work either.

The jellyfish problem was an easy one to solve; as a lowly plant operator on the Copperbelt I had made up enough solutions of flocculant to know what happened if the process is short-circuited. The dry white granules, varieties of a polymer called polyacrylamide, adhere together in a fearsomely sticky mass if wetted, which is incapable of being dissolved. To dissolve them

All poor together

in water they had to be trickled slowly, almost grain by grain, into the vortex of a mixer. Luckily Buckreef had spare mixers, and Steven briskly made plans to put them on the walkways across the thickeners so that they could dribble the slimy solution into the feed channel taking the pulp to the centre.

The grey anti-corrosion paint on the thickener had flaked off in places, revealing a silver surface. A weld running through it was also uniformly shiny; welds normally show the first sign of rust, and this thickener had been through at least two rainy seasons.

'This thickener ... it's not built of stainless steel, is it?'

'Yes.'

'And the drum filters? The piping?'

'I think so.'

'And the agitators?'

'I don't know.'

They at least turned out to be made of mild steel, which is the standard on gold plants and is about a quarter of the price of stainless. Yet the Buckreef plant had enough unnecessary stainless steel in it for perhaps two and a half million knives and forks.

The drum filters were not exactly a mystery like the static filters, but they were the first of their particular type I had seen. Instead of the filter cloth being wrapped securely around the three-metre drum (inside was a cat's cradle of pipes to suck the clear cyanide solution through), it was fed over a long roller, where the paste of ground-up rock—the cake—that was left behind could be scraped off. Göran and Steven had found that the cloth had a life of its own, slipping sideways off the roller and eventually off the drum as well, so that pulp, not solution, was sucked into the piping inside.

This was alarming; without the drum filters most of the gold in solution would be lost to the tailings dam. Not incidentally, so would the cyanide with it. The solution remaining after the tailings had settled out in the dam was discharged into the local stream. Suddenly the chance of accidental deaths from cyanide poisoning at Buckreef, which I had disdainfully dismissed, did not seem so remote after all.

Göran and Steven had tried everything, changing the filter cloths (three months to arrive), positioning workers 24 hours a day

A very good book

at the settings, modifying the operating cycle, using micrometers to make the adjustments and so on. Appeals to the manufacturers brought no better ideas. The only consolation (although for whom I was not sure) was that, as the drum filters had also been made of stainless steel, their scrap value could be millions of dollars.

Much later I discovered that this unique design had not worked anywhere else either. Why the middle of the Africa bush was selected for a trial I cannot know. But I can guess; if things went wrong the news would probably never leak out, and if it did the problem could be blamed on bad operation. If the equipment were successful there would be time to patent the modifications the user had made, and prepare a marketing plan before competitors could act. A surprising number of trials of new ideas are first undertaken in the wilder parts of the world. In the nature of things most of these fail, conveniently without publicity.

As it happened, drum filters were one of the pieces of equipment on which Göran had experience; iron ore mines have occasion to use them. At the time I didn't know that the drum filter design at Buckreef was a hopeless dud, but as we talked our way through the desperate rearguard actions they had taken, I could think of nothing that I would have done that they hadn't already. I shook my head, and Göran's face lightened momentarily. The expert was beaten as well.

Then we entered the realm of the static filters, and there, amidst the network of piping, memory returned. Like a modern brewery, I had thought. That was right, for a new one back in Salisbury had installed these units. I had glimpsed the plant a few years ago while negotiating to buy some of their old pumps. Polishers, that's what these filters were called there. They existed to take the last few, fine particles out of a solution (or a brew) that was already, in the evocative phrase of the old-timers, *gin-clear*. But at Buckreef the clay and jellyfish in the overflow from the thickeners and the slurry pouring into the exposed collection tubes of the drum filters was choking the static filters in minutes.

Well, I could get a clear thickener overflow, but there was nothing to be done with the drum filters. So here then, in the noon sun by the static filters of the Buckreef Mine mill in Tanzania, was

All poor together

the point at which my magic sequence of good luck stopped. The whole plant was so unsuitable, so badly designed for what was supposed to do, that it would be best to stop trying to fix it this very moment, and to go home never to return, with the modest success of having shown Steven and Göran how to use flocculants properly, leaving this disaster of a mine to fester on unseen in the depths of the forest.

Göran and Steven had been going on about the clay while I had been arriving at this conclusion.

'We don't know where it is coming from.'

'There is none in the ore; we've looked through the stockpile.'

'If there wasn't any clay the static filters might work.'

This was the final, unanswerable mystery; their defeat had come from a malign presence, as elusive as a witch doctor's curse.

'Have you checked underground?' I asked.

'Sven hasn't seen anything there.'

'We should check, don't you think?'

Eight years before and nearly three thousand kilometres southwards the performance of the mill at the Empress nickel mine had deteriorated badly. Although the crushing and grinding of nickel ore was done just as in Buckreef, the nickel in the pulp was concentrated using flotation. The big mill building—Empress handled a million tonnes a year, Buckreef, very optimistically, a tenth of that—was filled with ranks of long frothing tanks and this froth was collected, for it carried the metal. The recovery of nickel was normally about 85% but now it was falling to 65% and Empress mine was going bust.

The mill manager seemed incapable of correcting the situation, so he was fired, as was proper, and we—of the research and development department of this group of mines—were told to fix it. Providentially I came to talk in the bar of the mine club with the mine geologist, who had just acquired his master's degree on the strength of a description of the Empress ore body.

He was anxious to spout to somebody—anybody—about his achievement, using words like gabroic, peridotitic, pentlandite and metamorphism. Because my eyes glazed over at these slightly more slowly than the bellowing miners about us, from him I learnt

A very good book

that the Empress ore body was a bit like a stumpy carrot, with a core where the nickel was in the form of much finer particles.

More difficult to treat? Oh yes, quite probably, said the geologist. Why, the miners were now reaching that core, so the effects should be showing up in the mill. Was anything happening there?

This sort of thing is more common than it should be. On a much larger scale South Africa's first attempt at self-sufficiency in phosphatic fertilizers came badly unstuck in 1956, because the concentrate from the mine at Palaborwa consumed an uneconomic amount of sulphuric acid during its transformation into superphosphate. It was known that it would be difficult to concentrate that particular ore, but in the self-congratulation once an equivalent grade to imported Moroccan phosphate had been achieved, it seemed to have been forgotten that a concentrate from an igneous rock might have alkaline impurities. Only the fact that the mine was a state enterprise, meaning that politicians' reputations were at stake, kept it going until, two years later, a method for producing a clean concentrate was achieved.

So, back at Buckreef, Göran, Steven and I went first to call on the mine geologist. He was a lanky, sensitive Tanzanian called Lucas Mwambo, who had gained *his* M.Sc. in Rumania, where he had played football, his first love, and found himself the focus of racial abuse when he out-played his opponents.

Lucas produced a paper he had written describing what he found about the geology of Buckreef. Most of the gold, it appeared, was concentrated close to the side walls of the reef, but there was nothing there about clay. Yet neither Göran nor Steven had been underground on this, their mine, and it was clearly time that they should do so, with Sven, Lucas and myself.

There were, I remember, no spare cap lamps—we visitors had a torch belonging to Göran between the three of us. Nor were there any overalls, safety boots, lamp belts or gloves. Hard hats were obtained by borrowing them from a trio of underground workers languidly breaking rock on the stockpile. Sven, who was our guide, was properly kitted out with the right clothing, down to safety glasses, but he had brought it with him from Sweden.

All poor together

Stamico was responsible for dressing the miners, so the dearth of safety clothing should not have been a surprise. However, the Swedes had put in the underground section at Buckreef, and they had done well. There was a good, fast, hoist, a twin-compartment shaft, proper ventilation and pumping services and all the necessary development—drives, raises, cross-cuts—in place to enable the mine to produce two hundred tonnes a day for several years.

As I remember we did not go down in the hoist cage that first descent, but climbed down the emergency ladders at the side of the shaft to the first level, thirty metres below the surface. This was because the hoist did not stop there. It emerged that Buckreef was not a new mine after all, and the first level—'one level' in mining talk—had been worked out a score or more years ago, in colonial times.

We stumbled over the rusting rails and pipes in the drives there, looking at the damp walls. Plenty of clay here, because over the aeons rainwater had soaked through the rock surrounding the reef and transformed it into a soft, sticky substance. It was an unsafe place as well; there were several parts where the roof had collapsed. But nothing was being mined from this level; the gold had already been taken. The only inhabitants were scores of bats, who swept past us in rustling hordes and emerged suddenly from the shaft (or so we were told later) like a brief black cloud. The dank tunnels stank of their guano.

We returned with some relief to the shaft and recommenced our descent on the ladderway, inching down towards an area below of light and noise. From it drifted up the healthier tang of freshly broken rock and a whiff of the acrid gases from blasting. This was two level, and most of the ore on the surface stockpile had come from here. Three other levels lay below us, so the deepest was one hundred and fifty metres down.

At last we stepped off the ladderway into the shaft station, following Sven, who was looking about in a proprietorial manner. A small of miners was sitting on a bench, provided to keep people out of the way as they waited for the cage, and they stood up as he came into view. Mining is a dangerous activity, hence disciplined and hierarchical. Even the management titles echoed this. On

A very good book

surface the people in direct charge were known as shift foremen, down here as shift bosses. Shift foremen reported to a plant foreman, shift bosses to a mine captain. Buckreef's mine captain joined us here, another short, confident Machusa called Sidney.

Tubs of fresh ore were being pushed into place ready for hoisting, while an empty one was being rattled away by two miners along uneven rails into the darkness. As we followed them along the level, the burring noise of a compressed air borer—a jackhammer, it is called—became louder. Eventually the cocopan reached a row of chutes sticking out into the drive. A miner lifted a restraining plank and a torrent of rocks poured down to fill the tub.

The pushers—trammers they are called—bent to their work and started to shove the loaded thing back along the track, but Sidney gave a sharp command and they stopped. Sven gestured at the broken rocks shining wetly under our lamps.

'No clay there.'

The fresh ore was composed mainly of white, barren-looking quartz. So perhaps this was the buck reef. I had panned a sample of such quartz earlier while we were with Lucas, and was dismayed at the tiny tail it gave, just a few fine particles of gold. The rock looked clean enough, but then, quartz is not a clay-forming mineral. There were also pieces of dark greenish stone in the cocopan, finely striated, or schistose as the geologists call it. This was the country rock, some of which had broken off to dilute the gold-bearing quartz, but although this can form clay, given enough water and time, there was no sign of that here.

The gold mines of many countries, including Zimbabwe as well as Tanzania, are found in this type of rock. It is amongst the oldest on earth, and the geologists call the era when it was formed the Archaean. Prospectors look for greenstone schist before they start to look for gold.

I picked out a lump of greenstone; there would be little or no gold in it. 'Not too much waste,' said Sven. Good miners pride themselves on keeping the amount of valueless rock sent to the surface to a minimum. 'Yost a little goes up. But no clay. You see?'

'Okay. But I'd like to go into a stope.' He led the way to a rusty steel ladder, down which the noise of the jackhammer came

All poor together

clearly. We climbed up the raise for about fifteen metres. Here was another drive at the end of which, in a torrent of noise, a small group was grappling with a jackhammer as it drove into the wall. Sidney flicked the light of his cap lamp rapidly back and forth across their sweating faces and they stopped the drill, leaving a sudden silence broken only by the sharp hiss of compressed air.

Beyond the miners our lights beamed into misty space. This was the stope, the place where the ore was being mined. After the holes being drilled here were charged with explosive and ignited, the blast would throw the rock we were standing on out into this void and down to the chutes far below. I peered gingerly over the edge. The far wall of the stope was faintly visible about fifteen metres away across the gulf, and I could see down to the broken rock over the chutes, but the beam could not reach to the roof, twenty metres above. There was no support, no timber, no rock bolts; it was not needed. I knew that Sven was a competent, experienced hard-rock miner and he would have certainly had it put in if he judged it necessary. Not only was this buck reef a strong material, but so was the greenstone in the side walls. No clay here; we were far below the weak, damp stuff found at one level.

'Are there any cross-cuts around here?' I asked. Sidney gestured at the way we had come and we followed him back to resume our climb. As we did so the ear-splitting roar started up again, dwindling slowly behind us until, at the next sub-level, it had become just a thrumming in the rock.

Here a drive had been put out on either side. This was a very wide ore body, all of ten metres in places, and we went to the end, where the quartz met the greenstone. Geological contacts like this are often the most interesting places; at Buckreef, Lucas had told me, most of the gold was here. Lucas thought that the name meant simply that somebody had shot a buck on the outcrop, but the mine had been well christened, for the bulk of the white quartz in the mine was a true buck reef. It was too low grade to be worth much, but it was impossible to mine the high grade edges without taking the centre out as well. But no clay, not at this end of the cross cut.

Ore bodies are seldom straight up and down in the ground, but that at Buckreef dipped nearly vertically. Here we were looking at the greenstone of the footwall; that is, the side of the ore body resting on the country rock. We turned, a sombre group, and stumbled across the reef to the hanging wall contact. I scratched the greenstone schist; it was tough and hard. Water dripped briskly from above our head.

'See, no clay,' said Sven.

Where was that water coming from? I borrowed the torch from Göran and peered up.

'There!'

Now three lights were shining up on the contact. Moisture was dripping from a thin pale band, not more than half a centimetre across, sandwiched between the reef and the country rock. Water had found its way down weaknesses and cracks in the greenstone until it reached the impervious quartz, and then had drained down the hanging wall contact. Over geological time this trickle had turned the greenstone there into a thin layer of the same sort of mud we had seen on one level.

It's not too much,' said Sven, meaning that it was a very small thing to have caused all that trouble up there on surface.

Enough, I think.' I looked at Göran, who nodded. He was perceptibly more lively already. It would be hard now to blame him for something that, in effect, God had done.

* * * *

The clay had been invisible in the ore at surface because the blasting that broke out the reef caused it to disintegrate into tiny particles. Indeed, that was why there was so little waste rock coming up with the quartz; the weakness at the contact meant that the reef broke cleanly away at that point.

When we swept up back into daylight, we were bubbling over with good humour, even Sven. It was a cheerful party that sat down to a lunch of scrawny chicken and rice in the Swedish House. Nonetheless for me Buckreef had still been a defeat; honourable, yes—the cause of the problem had been found—but they were not going to produce gold at Buckreef any time soon, not with this collection of useless drum filters, clarifiers that were meant only for breweries and two vast thickeners, valuable only for the stainless steel they had been foolishly made of.

Or maybe not. I tried to remember the system they had used at the Renco mine, for years now the biggest underground gold producer in Zimbabwe. But what I was trying to recall was what happened at little Renco, its predecessor, during the war—our war, that is.

• • • •

Robert Kennedy was a mining engineer who had made a name for himself both as a very good rugby player and as a first-rate smallworker. A smallworker was what one called a miner—almost invariably in those days a European—who ran his own little gold mine. Not unassisted; he (or sometimes she; Robert Kennedy's mother was a smallworker) would have a labour force of between twenty and a hundred, depending on the size of the mine. However, this being Africa a smallworker was not just the boss, he was engineer, miner, metallurgist, geologist, assayer, medico and all the rest.

My father was not exactly all this, but he did have a third share in a smallworking and spent quite a bit of time there. This was a truly tiny affair, just an adit into a hillside and a stamp mill at its entrance. After the mill there was only a copper plate covered with mercury to catch the gold. On an average day the mill would treat about ten tonnes of ore and the mercury amalgam would collect perhaps an ounce of gold at a time when the price was only thirty-five dollars an ounce. Robert Kennedy was an order of magnitude above that, for he was running a fifty-tonne-a-day mine, hidden in the hills about eighty kilometres south of the ancient stone structures now called Great Zimbabwe. (At the time of which I am speaking they were still called the Zimbabwe ruins.)

His mine was called, for reasons I have forgotten, the Renco. It was certainly ancient, indeed it was the first mine—and probably the only one, too—that the traders saw when they came to barter for gold at Great Zimbabwe. If you look at the map of Africa, there is a marked indentation opposite Madagascar, which marks the site of a now-abandoned port called Sofala. The Portuguese captured it from the Arabs early in the sixteenth century, but the fort they built has been overcome by the encroaching sea, and only a few pieces of the walls remain, stranded like hulks far out on the mud flats at low tide.

That indentation makes Sofala the closest point to the gold-bearing highlands of the interior, and traders could make their way from there to the market at Great Zimbabwe in only a twenty-day march. Their route took them up the Busi River, over the divide to the Save River, and from that up the Mutirikwe River, which runs past Great Zimbabwe.

The Mutirikwe also passes Renco and to this day in the river valley below the mine stands a clump of date palms. These are an exotic species in Zimbabwe, and they can only be descendants of ancestors that sprung up four hundred years ago, from the date stones that gold buyers from the coast left there.

The Renco was something of an enigma. It is off the beaten track in a geological sense, lying well to the south of the nearest greenstones, on the edge of a vast structural feature running from Mozambique to South Africa that is called the Limpopo mobile belt. It will be understood that, since geologists deal in aeons, if it ever had been mobile it was a very long time ago. The point was that everything was fairly wrenched about in it.

Three major mining houses (Anglo American, what was then known as Gold-fields and what is now known as Gencor) had tried, and failed, either to elucidate the rich but confused geology or to crack the equally daunting metallurgy, and in about 1972 Robert Kennedy acquired the property. Along with what he inherited from his distinguished predecessors he created a mine and made some money out of it.

However, he knew that he could never do this for any length of time and that the massive, challenging gold resources beneath were beyond his capability to develop. Robert had taken his degree at the Camborne School of Mines in England, where an annual rugby match bearing his name is still played. One of his contemporaries there was an older man, Bill Rickards, an ex-Lancaster bomber captain and prisoner of war. Bill was my boss,

All poor together

or rather my boss's boss, and went with his chief geologist, F. C. ('Chick') Böhmke, to look at the Renco.

Robert Kennedy had little truck with conventional geology and, once underground, he produced a dowsing rod of copper wire, with which he proceeded to demonstrate to the visitors where they could expect to find gold if their company bought the mine.

The spectacle of two middle-aged mining executives and a veteran geologist carefully watching the twisting of a piece of copper by the light of their cap lamps was not quite as bizarre as it might seem. It had, after all, been mentioned, if not endorsed, by Agricola, the great patron of mining engineers, in 1556—

> There are many great contentions concerning the forked twig, for some say it is of the greatest use in discovering veins, and others deny it.

Chick Böhmke, I know, was decently sceptical of the technique. He had had an odd experience when Sir Val Duncan was making Rio Tinto into one of the world's great mining companies. He had been flown to a Greek island and had sat with Sir Val watching the behaviour of a pendulum being held over a small-scale geological map by the spoon-bending Israeli, Uri Geller. However, no major discoveries arose, probably because the pendulum insisted on locating minerals in the deeper parts of the oceans on a map whose scale was such that a centimetre represented several kilometres.

Whatever his scepticism, Bill Rickards bought Renco mine. So it was that my modest department was faced with solving a metallurgical challenge that had defeated three major South African mining houses. For reasons that are rather unimportant at this distance, I ended up on night shift on the plant at Renco, doing tests. Because it was the middle of the war I worked with a revolver in my belt and was trailed by two guards, Police Reservists like myself. Indeed the little mine was in a state of siege, with most supply convoys being ambushed and the occasional shot flying in our general direction.

There was a happy ending, for eventually we did crack our metallurgical problem as did the geologists theirs, hence today's

A very good book

big mine. This was largely paid for by Rio Tinto in London, who had been uncomfortably aware that their erstwhile subsidiary had prospered during the Rhodesian rebellion to the point that the parent company had become minority shareholders. By paying for Renco the parent company enabled the *status quo* to be recovered. As a postscript, that may not have been a good thing. Rio Tinto is arguably the world's biggest mining company (if you count Anglo American and De Beers as separate organizations), and because Zimbabwe has the sort of geology suitable only for smallish gold mines (Renco itself is a medium-sized mine by world standards), the company has not opened any new ones since the end of the rebellion.

All this digression leads to the point that Robert Kennedy's little Renco had a unique flowsheet. Practice in South Africa and Australia used drum filters to separate liquid and solids, just as at Buckreef. However, his plant used a system popular in North America, counter-current decantation, involving three or more thickeners. Even now I cannot say which is the better; perhaps on balance the more complicated drum filters, where at least if anything goes wrong you don't have to gallop across half an hectare of thickeners to fix it.

Robert Kennedy was killed in an ambush later in the war, but not before he had launched a gold mine called the Grandeur on the local stock exchange. This foundered comprehensively, despite its impressive name, and gave what is now the Zimbabwe exchange such a fright that no new gold mine flotation has been permitted on it since that time. This was, and is, a much greater reverse to mining in Africa than it might appear; since that time only big blue-chip mining companies have been allowed to raise money on the Zimbabwe stock exchange. 'Juniors', of the type that have thrived on the Canadian and Australian exchanges and which discover many of the resources that the major companies have ultimately developed, have not been made welcome. For the time being, though, we are sitting around the scratched wooden table in the Swedish House, chewing bony chicken, and I am thinking about those two big thickeners of, great heavens! stainless steel.

• • • •

Steven and I stood again on the platform above the agitators and looked down on the silent equipment beneath. This time Göran was with us as well, and gesticulating with some enthusiasm at the modifications we would need to make. In his hand was the piece of paper on which we had sketched out the changes. Luckily, one of the few things in common that iron ore plants have with gold mills is counter-current decantation. Inevitably, this system is conveniently abbreviated into CCD.

The two thickeners in the circuit had been intended to play a very different role from the one we were planning. One increased the thickness of the pulp from the grinding mills so that the volume was less and so the size of the agitators could be reduced. Since there was a ridiculous amount of agitator capacity anyway, it could be pressed into other uses without difficulty. The second was there to increase the density of the pulp leaving the agitators yet further, so that the drum filters had a minimum of solution to extract. As the drum filters were to be abandoned for CCD separation, this thickener too could be pressed into other duties.

(Envisaging the CCD circuit as a mental exercise is difficult, hence the hotly argued-over diagram Göran was holding. Unless you are of a technical bent, skip the rest of this paragraph.) The pulp with gold in solution coming from the agitators went into the first thickener, which was well doused with flocculant. The overflow from this would go a shut-down agitator whose floor was to be hastily converted to a sand filter. From there the clear solution would be sent for polishing to the Static Filters. The underflow from the first thickener would be mixed with fresh water in and sent to the second thickener. The underflow from that would go to the tailings and the overflow sent to the pulp flowing into the first thickener, whence it would contribute to the volume of pregnant solution. Trust me.

With only two thickeners this was a very inefficient arrangement for separating the pregnant solution from the cyanided ore. About 30% of the gold already in solution would be lost to the tailings, along with a lot of cyanide, and something would have to be done about that. Robert Kennedy's plant had three thickeners, which reduced the loss to about 8%, and big plants typically have four,

A very good book

which drops it to about 2%. But at Buckreef a 70% recovery would be vastly better than zero.

To balance this circuit it would be necessary to send cyanide solution from which the gold had been removed (called barren solution, naturally) back to the grinding mills. So we also abandoned the elaborate system that had been installed to make up and feed concentrated cyanide solution to the agitators (now *that* was a dangerous liquid) and put briquettes of cyanide by hand—gloved hand—on to the mill feed belt. This is both much safer and usually gives splendid gold extraction in the grinding mills, even before the pulp reaches the agitators.

The above description has been put in largely for professional interest, if you can call it that. For those who are just hoping for a good read, what matters is that once we had agreed on the arrangements, it was largely a matter of rearranging the connections of a number of big rubber pipes.

I was woken the next morning by a new noise rising above the throb of the generators; the fierce, steady roar of the grinding mills. I shaved hastily and walked across to the mine. *'Habare?'* called the children at the entrance. *'Mzuri!'* I called back, pleased with myself for remembering the reply, but they still ran giggling away. The boom was briskly raised to let me in. At the plant Steven and Göran were standing on the first thickener looking pensively at the surface. Behind them the solution from the new flocculant addition system was dribbling into the pulp running in from the grinding mills. I had gone to bed late; they had hardly slept at all.

The flocculant would take a measurable time to work, so the thickener was recycling to start with. But two hours later, by eight o'clock, the solution coming over the lip of the thickener was clear enough to send it to the improvised sand clarifier in one of the agitators, and the second thickener was started up.

We climbed down and across to the base of the agitator, whose bottom now had a sequence of bricks, sacking and thirty centimetres or so of washed sand filling it. We could hear the pregnant solution splashing down inside the high tank into an empty drum, pierced with many holes, placed on the surface of the sand. Its function was to break the impact of the fall, so that the filter bed did not get eroded away.

Steven opened the drain valve at the base and we followed him along the newly installed piping to the static filters. Hekki was with us now; he had driven in yesterday evening, and was to take me back to Mwanza later on today on the first stage of my return journey. He had brought news; Mr Vergympf had been found and Pat Carter had arranged for a helicopter to bring them both to Buckreef from Mwanza this morning.

So I was as anxious as any of the tense group watching the line of pressure gauges. They quivered into life as the pumps started up and began to force the clarified pregnant solution through the static filters. Would the gauges sweep around to their limit as clay particles choked the filters? Up to now they had done that within a few minutes of every start-up. The needles inched upwards, slower and slower, and, then, heaven be praised, stopped and held steady. Göran went to the discharge pipe leading from the filters and opened a drain valve. The pregnant solution gushed vigorously out and he caught some in a glass beaker. He held the liquid up to the bright morning sunlight and smiled for the first time.

'Gin clear,' he said.

● ● ● ●

We had a cheerful late breakfast and went back to the plant, which by now seemed to be working pretty well as planned. (However, and perhaps by this stage not unexpectedly, it turned out that there were major difficulties in making the, as yet, unused gold precipitation section work properly. But that was later.) Hekki kept looking at his watch; we had to leave by one o'clock if we were to be sure of catching the last ferry, and still there was no sign of the two important visitors.

In the end we had to say goodbye and go. We pulled away from the Swedish House, out on to the main track and drove past the low houses scattered in the forest, down the dip and across the stream that marked the mine boundary and up the other side. As we did so, a familiar brisk whap-whap-whap was superimposed on the grinding of the Land Rover. I glanced back. A small yellow helicopter came racing over the forest, hovered, and sank down out of sight towards the mine. Hekki looked at me and shrugged, and we drove on.

* * * *

That was, of course, not the end. I eventually went back to Buckreef twenty-two times, becoming partially immunized to the chaotic ways of Tanzania, and with my Swahili slowly returning after twenty years of disuse. I was there when Göran and company left after their contracts expired. None of them wanted to stay on, but in the way of things their relief was mixed with a little nostalgia when they bounced away from the Swedish House for the last time. The fresh team of Swedes who replaced them did not stay long, and when they went the whole mine was run by the Tanzanians, a much bigger achievement for the aid business than might be supposed. Thanks to continuing desperate improvisation, the plant was kept working. At one stage we stripped it of its handrails to provide piping to a vacuum tank, while the vacuum itself was obtained by connecting these to the air intake of an unused compressor.

As a result, modest amounts of gold continued to be sent out at erratic intervals. When it arrived in Dar-es-Salaam it was carried in a police car, siren going, with an escort of outriders, through red lights, directly to the Central Bank's vaults. Back in Buckreef I was now the only occupant of the Swedish House, and, I suppose, the sole European for a radius of a hundred kilometres or more.

It was all a waste. Ten years, ten million dollars, all of Göran and Steve's valiant efforts and those of a good number of other unsung heroes, were for nothing. Worse, for less than nothing. As I have suggested at the beginning of this book, it would have been better by far just to give the money away to people who looked as if they might do something with it. Buckreef did not just squander Swedish taxpayer money, it also wasted that of Tanzania and, worst of all, damaged the confidence and self-esteem of the Tanzanians.

Not one problem but a whole hierarchy of them caused this

All poor together

failure. At the bottom, and the only one that offered serious hope of being overcome, was the immediate difficulty I had been summoned for—to fix the plant. This, then, was the only one that made some progress, although the arrangements that we cobbled together to produce some gold were terribly inefficient. To dispose of this matter here, at the time it seemed to me that the best move would be to change the flowsheet to a new type that used activated carbon granules in the pulp to adsorb the gold directly from the cyanide solution. These would then be simply screened out of the pulp and the gold extracted. This 'carbon-in-pulp' system does away with the need for thickeners and filters, and while it was still something of a novelty then, it has since become the standard for gold plants.

In any event I was wrong; the next technical problem up the ladder was much more fundamental. It was to do with the geology. Buckreef, like Renco, like my father's mine, treated a class of gold ore known as 'hydrothermal'. With a certain diffidence I present the theory behind what is thought to have occurred to create the ore in such cases. Diffidence because, in the presence of a sceptical layman, it all sounds rather implausible. However, both the attraction and the foundation of much geology is Sherlock Holmes and his 'whenever you have eliminated the impossible, whatever remains, *however improbable*, must be the truth!'

The greenstone schists of such mines are, as I have said, very ancient rocks, and most of the gold in them is thought to have come from nearby, more recent, granite intrusions. The hydrothermal theory envisages very hot, rather salty water being squirted out of the cooling and contracting granites into cracks or weak spots in the surrounding greenstones. Much silica had dissolved in these solutions, as had gold and other minerals, including iron and sulphur. The water eventually evaporated, leaving the silica, which became quartz on solidifying, while the iron and sulphur formed various types of iron pyrites, also known as fool's gold. Real gold is often found with, or in, pyrites, and once it is there, cyanide won't get it out. There were pyrites at Buckreef, clustered, like the gold, at the contact between the quartz and the greenstone schist.

A very good book

Now Lucas, the Tanzanian geologist at the mine, knew this, but not, evidently, the string of expatriates who designed it. Or perhaps it was my erstwhile brothel companion Pat Carter who led them into making the mistake. Only one test was undertaken on the ore, and that on a single piece that was smuggled down to enemy territory in South Africa.

It would have been possible to carry out a string of comprehensive tests in Tanzania; there was a decaying government laboratory up-country in Dodoma that could have, in theory, undertaken the work. But I think that Pat Carter secretly trusted his own countrymen less than the South Africans on this matter, and given the circumstances I would have felt the same. However, I would have at least ensured that the sample that went to Johannesburg was a representative one.

The piece of reef that did go there turned out to have very little gold in it, but what there was came out readily enough when cyanided. So the sample suggested that no problems should be experienced in a conventional cyanidation circuit.

We have already met the ore that must have been sent to Johannesburg. Yet even a fairly knowledgeable mining person might be misled into thinking that the lump of nearly barren white quartz on his or her bench was from the ore body, and that the sample just happened to be very low grade. Gold distribution is almost always very erratic. Yet we know that Buckreef was truly named, and that the great bulk of the ore had intrinsically very little gold and even less pyrite.

What the laboratory should have been testing was the rich material at the contact between the quartz and greenstone, where it would have been found that while this was high grade, most of the gold was locked in various forms of pyrites, and would not leach out with cyanide. Why, such a sample might even have included some of the clay at the hanging wall contact, making the researchers think twice before recommending a conventional filter circuit.

Or perhaps not. The gold on the Witwatersrand—still by far the biggest source in the world—is also mainly in a type of white quartz, and the metallurgist testing Buckreef ore there may have

All poor together

known little of hydrothermal gold. For the 'Rand' is probably the result of a vast, very ancient alluvial deposit, a three-hundred-kilometre-long series of pebbly river deltas edging an ancient sea. (I say probably, for, again, there is at least one other theory about its origin, but the one I give here has at least the merit of some plausibility to the layman.)

This Witwatersrand gold may have originated far away in a hydrothermal source, and much twentieth-century history arose from this possibility. This will be told in due course; enough to note that aeons ago the gold was washed down to the shoreline, where it had more aeons of exposure to air and water before geological changes buried it deep again. There would have been none of the original pyrites left by the time the quartz surrounding it began to turn back into rock once more, and so, unlike the Buckreef ore, Witwatersrand gold is what is termed 'free milling'.

If a representative sample had been sent from Buckreef, no doubt the observation would have come back that this was nothing like as simple to treat as Witwatersrand ore, and that flotation, the same process that was used at Empress nickel mine, had to be used. There, you will remember, it was driven home to me that the place to start looking for explanations, if not solutions, for a problem in the surface plant is underground, with the geologists.

If Buckreef had produced a flotation concentrate of gold-rich pyrite, it could have been roasted to break down the pyrites by oxidizing away the sulphur. After that the gold in the rust-red calcine remaining could have been cyanided out satisfactorily. That, then, should have been the plant flowsheet.

But it was never changed. The rushed, temporary arrangement of pumps and pipes that we had thought up over lunch that day in May 1982 remained in use until the mine finally ground to a bankrupt halt six years later. It was not that cash was the problem, for during that time perhaps a million dollars of aid money was spent on a huge Finnish generator that was as unnecessary as it was expensive. In addition Swedish aid funds continued to trickle in to avert the repeated crises with urgently needed spare parts and chemicals that occurred.

No (and here we come to the third, most obdurate, problem),

A very good book

Buckreef's quandary was not money, it was that it was a government project, immutable in detail once the budget had been approved and the decision to proceed had been made. The discovery of the difficult—the technical term is 'refractory'—nature of the ore was not to be allowed to stand in the way of the process. The mine had its committed champions in Dar-es-Salaam. Stamico and the Ministry of Minerals had made it part of their development policy. To admit, now, that major, expensive, changes were needed for it to work, after ten years of headlines in the (exclusively government-controlled) press saying things like 'Buckreef Set to Transform Tanzania's Mining Industry' was unthinkable.

That I was to learn, and learn well, in the future. I was also to learn much more about the aid business in Africa, to the point where now I believe that even if all three levels of difficulty had been surmounted, the mine would still have been doomed to economic failure. But for now I was bumping back over the long road to the Kabanga ferry, watching the sun sink gracefully into Lake Victoria, enjoying a hot shower in Mwanza and flying off the next afternoon in a packed Air Tanzania Fokker Friendship (it was overbooked; somebody's child was strapped on my lap for take-off and landing) to Dar-es-Salaam.

The next day Hekki took me to the Stamico headquarters, a grimy and run-down converted apartment block, besieged with the rusting wrecks of what had once been its transport fleet. In the cramped office (which must have once been somebody's bedroom), that he shared with another Finn, he copied out a list of the recommendations I was planning to make in my report. On the desk was a familiar, well-thumbed book with a dark green cover. I have my own copy. It had, Hekki and Pentti both explained, been invaluable for designing the plant.

'A very good book,' said the latter. I had to agree. For what it was, the very best. It was called *Gold Metallurgy on the Witwatersrand*.

Chapter Three

THE STUDIOUS SHEPHERDS

In my early years I read very hard. It is a sad reflection, but a true one, that I knew almost as much at eighteen as I do now.
—Samuel Johnson, 20 July 1763, from Boswell's *Life of Johnson*.

A little learning is a dangerous thing, but Hekki and Pentti both had much more than a little. Indeed, they were only the most recent in a whole series of ostensibly well-qualified aid workers who had inhabited the little Buckreef project office. The accumulation of egregious and/or expensive mistakes that I had found at the mine—the wrong flowsheet, the unnecessary stainless steel plant, the unworkable drum filters, the choked static filters, the lack of a proper cyanide antidote and so on—generally long predated their own involvement; after all, the mine had taken ten years to build. But they had not done anything to fix them. Had they even known?

I was not so foolish as to jeopardize my own future by saying this, but it did start me thinking, particularly as I was none too confident of my own mastery of the trade. Getting the plant started was a great coup (as you will hear I was to work off and on for years in Tanzania on the strength of it), but I knew only too well that there was no magic in the process. It was more of a case of a one-eyed man in the land of the blind.

Yet if what I had seen was typical of the sort of thing that happened with technology transfer (and already, at that early stage of initiation, a couple of conversations with other aid workers in a bar in Mwanza the day before had given me an inkling that Buckreef was not unique) then 'development assistance' might even be doing more harm than good.

The long-term answer lay surely in better education for Tanzanians, so that they were not so dependent on foreigners to do

The studious shepherds

their thinking for them. Education with a practical bent, of course, but concentrating on the three Rs and encouraging an entrepreneurial outlook.

I was not alone in thinking that education was the key. It had been, I found, an article of faith in the aid industry for many years that education was a prerequisite for development. The arguments were impressive. No country, no matter where, had succeeded in making its population as a whole wealthier when most were illiterate. When wealth was thrust upon such countries—as with many oil producers—all that happened was that the already wealthy (and literate) got much wealthier. The poor stayed poor, even when money was thrown in their direction; neat fielding by the elites usually intercepted it.

That much was clear. In addition many of the politicians of the newly independent African countries, Julius Nyerere himself, for example, had been schoolteachers. So Tanzania was investing a major part of its budget (12% in 1981) in education and effectively all boys (and most girls) were attending primary school.

Yet it seemed to have been wasted. A year or so later we were trying to find candidates for shift foreman jobs at Buckreef, for which an ability to complete the log sheets and reports was needed. Not in English but in Swahili, because Swahili was the language they had all been taught. Not *in* which they had been taught, because for upcountry Tanzania Swahili is not an indigenous language; it is a lingua franca of Arabic and coastal Bantu dialects whose provenance is East Africa's harbours. However, it was designated as the official language of the country, stemming from Nyerere's determination to put a distance between his regime and that of the British colonialists.

Now the primary school at Buckreef (there was no secondary school) was no great shakes. At that time it was a series of long huts of grass near the road, with gaps for windows through which the bowed backs of many studying children could be glimpsed. Nor was the academic record good; none of its products at the time of which I am speaking had managed to qualify to enter high school. But you might have thought that four years of education might have produced a minimum literacy.

Perhaps it did. What is certain, however, is that none of the candidates could qualify for writing up a simple record of events during their shift, except for two, whom we gratefully recruited. They had to be fired a little later, as they turned out to be illegal immigrants from Kenya, where they had been educated at private schools.

This is perhaps a good point to warn readers that the rest of this chapter, like some others ahead, is given over to a rant, in this case about education in Africa; what has gone wrong and why, and the attempts to cover up this failure (or ignore it). If you are reading this somewhat schizoid book for fun rather than interest, skip. (You can tell when to do so in future from the swarming of endnote references and a certain impersonality creeping into the tone.)

The lesson from Buckreef seemed to be that for most children in the country, the effect of four years of mediocre education, followed by a life of goat-herding in an environment where there were no books or newspapers, and only dim firelight to read by after dark, was not perceptibly better than no schooling at all. Indeed, because of the four years of goat-herding time lost, coupled with the wasted teaching hours of teachers and the wasted office time of administrators, the overall effect of attempted universal primary education for Tanzania must have been negative at that time. In economic terms, education in Tanzania had very high opportunity costs, particularly for the 80% or more of the population whose careers would be forged in the goat-herding sector of the economy.

This anecdotal evidence has, surprisingly, been born out by research. Every year the World Bank brings out a solid volume called the World Development Report, usually focusing on some particular aspect of development. The 1990 copy dealt with poverty, and gave, first, the heartening table opposite.[2]

This, the report said, underscores the importance of investing in education. But then it spoilt the effect by saying that sub-Saharan Africa is an exception, and gave the graph opposite.

As you can see, this indicated that putting money into education in Africa actually had a negative effect on a country's wealth, which tied in with what I had seen.

The studious shepherds

AVERAGE SOCIAL RETURNS TO EDUCATION %			
Region	Primary Education	Secondary Education	Higher Education
Sub-Saharan Africa	26	17	13
Asia	27	15	13
Latin America	26	18	16

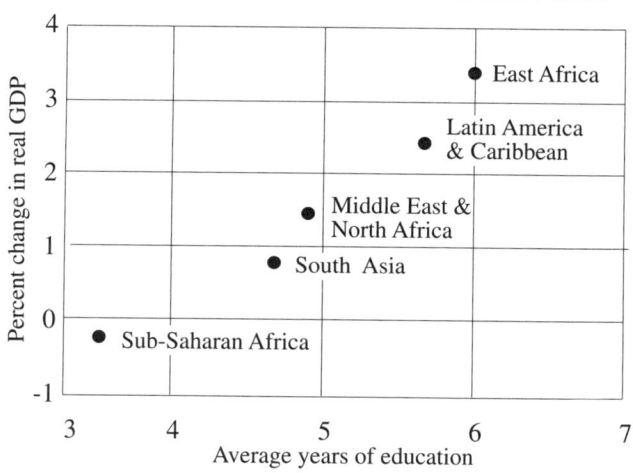

EFFECT OF A YEAR'S EXTRA SCHOOLING

So why do 'social returns' show a plus for education while country wealth in the form of the effect on gross domestic product show a negative? What are 'social returns' anyway?

Simply put, these are the estimated net returns to individuals (as a present value of his or her lifetime stream of income) after the 'private' benefits they get from education have been adjusted to allow for their share of the public sector expenditures on education.

It is one of those concepts that becomes fiendishly complex once you get into it. There is endless room for argument as to what should be put on either side of this private/state equation. Increased state benefits from the higher taxes due from better educated, and

All poor together

hence presumably better paid, people? The increased health, and lower health care costs, that have been shown to be associated with more education? The greater cost to the state of the increased education that, as has been repeatedly shown, the offspring of the well-educated individual will require?

But this does answer the question of why it is possible for education in Africa to look good in social return terms and terrible, possibly even malign, when correlated against its effect on the 'real GDP'. The social returns method is a fraud; it is stiff with assumptions about the pattern of future earnings, principally that they will grow. In Africa, people, educated or not, often have no future earnings flow, and those fortunate enough to be employed have been getting poorer. Of course, if you do manage to land a job (and here connections through the extended family may well be your sole hope), your qualifications will help, so that even for Tanzania there is a correlation between them and your income. Unfortunately, for most of the inhabitants this is a purely academic relationship.

However, the credo that Education is Essential was not to be lightly abandoned. In 1990 the World Bank made a commitment to double expenditure on education and health in sub-Saharan Africa so that it would reach 8–10% of country GDPs.[3] Something was achieved; between 1980 and 1995 the proportion of GDP spent on education went up from 4.1% to 5.3%, over a period when GDPs were broadly constant. However, during that period private per capita consumption shrank by 1.8%[4] and the average teacher's salary in sub-Saharan Africa fell by at least a third in real terms.[5]

Some of this was because those countries in Africa—now a big majority—whose economies have deteriorated far beyond their capacity to repay their debts, have been forced by their creditors to turn to the International Monetary Fund (IMF) for disciplining. The strict financial regimes imposed, called structural adjustment programmes (SAPs), have not helped education. The IMF's own figures show that in African countries with SAPs, the per-capita expenditure on education declined by 0.7% a year on average between 1986 and 1996.[6] From my own casual conversations with

The studious shepherds

IMF officials, usually because we were on adjoining aircraft seats, it seems that they believe the earlier attempts in Africa to provide universal primary education were unsustainable in situations where, on average, about 45% of the population is below the age of fifteen.

Indeed, the record of my own country, Zimbabwe, is a fine example. Our own ex-schoolteacher, Robert Mugabe, inherited, by African standards, a thriving business. Despite the war and United Nations sanctions, 85% of the population were at primary school (or had been; I saw a number of ruined schools in the last years of the war, some of them destroyed by what was, at the time, our side). While between 1965 and 1985 the average primary enrolment in sub-Saharan Africa climbed from about 30% to about 60%, in newly independent Zimbabwe it rose to over 120% of the expected number of pupils, as many children above (and some below) school-going age came into the classrooms.

In numerical terms, while at the end of the war in 1979 there were about 800,000 children in primary school, with the advent of peace this jumped to 1.2 million. However, because of the population explosion, the demand for free education is a moving target, and fifteen years later there was twice that number in the classrooms.

Entry to post-primary education had been restricted by the previous white regimes to the brightest, on the grounds that the country could not afford to do more. This was obviously not acceptable to any ex-schoolteacher, particularly one who was a Head of State, and in 1980 Robert Mugabe's government undertook to provide four years of secondary education to all primary school-leavers who had passed a simple graduation test. The result was that first-year secondary school enrolment leapt from 22,000 in 1980 to 153,000 five years later.

Such increases led to an unprecedented programme of school construction; in the first few years of independence Zimbabwe's foreign debt doubled, largely because of this. By 1996 there were 4,659 primary schools, compared with 2,401 in 1979, while the number of secondary schools available to students also doubled.

But by then they were giving a lousy education. Like just about every other country in sub-Saharan Africa, Zimbabwe had spent beyond its means and had been forced to go crawling to the IMF, so the education budget was shrinking. The expenditure per pupil had dropped from about U.S.$200 a head in 1989 to about U.S.$129 in 1996.[7] The primary school class size, which, by a frenetic expansion of the teacher-training programme, had been brought down to about 33 in the mid-80s, was up to 40. Those dollars themselves had lost value over that period; using the U.S. consumer price index, in constant 1996 dollars the ratio is more like $248 to $129. This was still pretty good; the average for all students in Africa, primary, secondary and tertiary, was U.S.$89 in 1990.[8] At the same time, and on the same basis, the developed countries were spending over U.S.$2,000 per student annually.

So now we have a situation in Zimbabwe where there are about 300,000 children leaving school annually. It is thought that only about 10% of them are able to find work in the sort of jobs that offer a career, a pension, deducts income tax and that sort of thing. The rest, if they earn anything at all, often do it by selling individual pieces of fruit, sweets and cigarettes from cardboard boxes at street corners.

The *Institutio Medio de Geologia e Mines* at Tete in Mozambique might exemplify all that is futile about education in Africa. Not that the education itself was bad in this case. I had come there (it was a United Nations evaluation mission) with a full set of prejudices about the Cubans who staffed it, only to find that they, and their Mozambican counterparts, were doing a very good job of training geological technicians. However, they were not, despite the name, training miners; these were geological experts.

Unfortunately, Mozambique at the time (1991) had nothing like enough jobs for the thirty or so geological specialists the school was producing each year. Further, the low-key but savage insurgency kept up by the Renamo *bandidos armadas* against the government (which really meant the civil population) had discouraged any geologizing more direct than airborne geophysics. This, too, was hazardous; most such work had stopped following the shooting down (probably by jumpy government troops) of a DC3 being used for this purpose a year or so before. Even now, in

The studious shepherds

the piping days of peace, there is probably employment for only about eighty such technicians in the country. In Zimbabwe next door, with much better geology and a score of exploration programmes, there are probably not more than a hundred or so such jobs.

The selection of students for the *Institutio* was based on a quota system from the various provinces. No attempt was made to ascertain their suitability for, or interest in, the course through a selection board. However, as acceptance meant the deferment of military service, it appeared that, while the initial candidates were eminently suitable (the Cubans had a say in their selection), succeeding groups were composed largely of those with no special eligibility beyond the right connections.

What Mozambique did need and still does (along with almost every other country in Africa), is individuals with a range of practical mining skills plus some financial knowledge who can put these to use in developing the numerous mineral resources suitable for small, simple mines. Why such mines do not exist in Africa outside Zimbabwe is a subject dealt with later but, as with Buckreef and so much else that might be done on the continent, there is not one but a whole edifice of reasons why this will not happen in Mozambique.

Firstly, and most easy to rectify, the Cubans were good geologists, but they freely admitted to knowing almost nothing about mining (Cuba has only a scattering of mines), and certainly nothing about the business aspects of mining. This could be remedied; Brazilians, for example, might have been recruited to fill this gap.

Trained, the potential smallworkers would need equipment. Here there was another problem, as at the time there was almost nothing, not even shovels, made locally. More sophisticated, but still simple, items like cocopans and rails, were just a dream, and the nearest they would get to the jackhammers and pumps needed were some elderly Russian models at a decrepit coal mine nearby.

Finally, if this were not discouragement enough, nobody, not even the recipients of this training, could acquire a mineable deposit through purchase; only the State and (oddly enough) foreign investors could do that. Locals would have the daunting prospect of having to set out to find their own. Yet, in a country where private property was outlawed (and property rights are still a disastrously unclear matter in Mozambique at the time of writing), any discovery they made would not belong to them except on sufferance. If they found something really good, the State would offer it to a large foreign company for exploitation. Mozambicans had no chance of benefiting from the market place for mineral rights that is the mainstay of all countries with a well-developed mining sector.

They would have one opportunity, however, that was denied to most of their fellow-countrymen. Their skills would be internationally marketable, no matter that they were not of a high level. Small mines everywhere are short of the type of all-rounders that the *Institutio* could have provided.

The loss of the best and brightest to more developed countries is a major factor in the marginalization of African education. One study found that, while Africans are only small contributors to immigration to the United States (only 128,000 out of a total of 7,000,000 during the period covered), 95,000, or 74%, of these had a tertiary education.[9] The United States, incidentally, was found to account for about 54% of the brain drain from developing countries, so another depressing estimate—that there are over 100,000 African professionals now living and working permanently in developed nations—seems to be of about the right order.

To return to Tanzania, a dozen years after my own discovery of the inadequacy of the schooling there, these disappointing results from the alleged panacea of universal primary education had been noted at the top. For the more worldly breed of politician that had succeeded the *Mwalimu*, schooling was clearly a poor use of government money. Anyway, Tanzania was bust, and had to cut costs to meet the targets for its own structural adjustment programme. So education's share of a diminished budget was down to 3.5%, and only 70% of children were going to primary school. The latest (1995) figures suggest (I say 'suggest', because the drop seems to have been overly precipitous) that by then only 48% of children of primary school-going age were enrolled.

The studious shepherds

But it may well be true. One of the commonest fiddles in forecasting the outcome of development is to use a time series that gives the required result when extrapolated. For instance, the 1965–85 trend line for school enrolment in sub-Saharan Africa extrapolates to a figure of 86% in the year 2000. However, the 1980–85 trend only gives you 46%.[10] And that, says the 1998 UNESCO World Education Report, is about what it is going to be.

Tanzania still prohibits for-profit primary schooling, but in practice it is becoming too expensive for parents to send their children to school anyway. The children are having to bring their own pencils, books and paper, chalk is sometimes absent and class sizes have swollen to a point where 60 is about the minimum. I know this, not just from my own experience, but because the *Mwalimu* himself admitted as much.[11]

Ironically, as if to show that the cynical actions of the politicians were justified, Tanzania's finances perked up during this time of falling school enrolments and deteriorating educational quality. In fact this was due to the gradual sloughing off in this period of the socialist straight jacket in which Nyerere had imprisoned the economy. Of course, some credit might be given to structural adjustment.

This need not deflect from the principal conclusion that Tanzania and other countries in Africa forced me to make about education there. It was largely a waste of money and (just as important) time. Cornering a young individual into spending years learning stuff, most of which would be never used and much totally forgotten, was no way to alleviate poverty. Rather the opposite. But why was this happening? Why was education at best a neutral force for economic growth in sub-Saharan Africa and a very positive one elsewhere?

No answers have been coming so far from official sources of development assistance, apart from the conventional response of asking for more money. The previously mentioned UNESCO 1998 World Education Report was typically sententious and unhelpful. 'If education is to help the poor to lift themselves out of poverty,' it said, 'then in the poorest countries education itself needs first to be lifted out of poverty.' It did not say why the earlier attempts to improve

All poor together

education in Africa by throwing money at it had not worked.

From Buckreef and elsewhere, the answer seemed plain. If what was learnt did not approximate to what was needed, then apart from the mental discipline it gave there was no point in teaching it. You may as well teach Latin, since the 'mental discipline' argument was the one its protagonists always used.

The corollary to this is that students who are likely to spend their lives in primitive subsistence agriculture, largely outside the cash economy, need very little education, and that in practical matters, such as farming practice and public and reproductive health. Beyond that, well, leaving school is not the end of learning, and adult education has been shown to be useful in these circumstances. Zimbabwe's evening classes only reach a tiny proportion of the population (about 16,300 in 1995) but the demand has made this the most rapidly growing part of the educational system, with enrolment up by over 300% between 1980 and 1988.[12]

So, education in Africa, a very conservative business, was locked into teaching a lot of irrelevant stuff. But this surely does not account for education itself becoming largely irrelevant to development in Africa. No, there are two other major reasons.

First, and most obviously, the population explosion has meant that the systems are swamped. For example, by 1996 in Zimbabwe there were about 3.25 million children in primary and secondary schools. Nobody knows for certain anymore what the country's total population is because of arguments over the real death toll from AIDS (as we shall see, one actuarial source estimates it to be about a million), but by extrapolation from the 1992 census the figure should have been about 11.5 million. So, roughly 28% of the population are at school at any one time. The comparable figure for the developed world is about 12%.

This is the big story, but it is not the whole story. Other countries have managed to educate a growing population to the point where an educated citizenry has made sensible choices about family size (smaller) and education (more). The Singapore government spends less per capita on education than Colombia and Rumania, but the products of their system have much higher maths test scores.[13]

The rest of the problem arises from human nature. A number of studies have demonstrated that services provided by the government, such as health, have tended to benefit the rich, not the poor. Recently it was shown from household surveys that, on average, the poorest fifth of the population of a sample of nine African countries got under 13% of the money spent on education by the government, while the richest fifth got over 32%.[14]

I can quote no references to prove the theory that follows, but a knowledge of humanity plus the experience of parenthood suggests that what happens is that the distribution of educational assets is heavily skewed towards the cities, where the richer sort of people tend to live. Here, as one can often see, most children wear uniforms, the schools have enough desks and chairs, there are some worn textbooks, and there is chalk to use on the faded and shrieking blackboard. Here, too, the monthly pay packet arrives more regularly than upcountry. Perhaps, too, textbooks are handed out without the need for an unofficial payment to the teacher.

Less obviously, this is where the best teachers are. That anybody living in the clean and quiet forest glades at Buckreef would want to be transferred to the noisy and befouled environs of Dar-es-Salaam may be incomprehensible, but it was the fierce ambition of every Tanzanian there, including, I am sure, the staff at the miserable primary school the mine possessed.

Even less obviously, here is yet another arena where Africa's equivalent of the old-boy net comes into play. Senior civil servants in the Ministry of Education are well placed to extend favours to teaching relatives in the matter of where they want to work. They would be failing in their primary cultural responsibility if they did not.

The bias in educational subsidies gets progressively more skewed as the level of teaching increases. In the early 1990s in Ghana the richest 20% of households acquired 45% of tertiary educational subsidies; in Malawi it was 59%, while in erstwhile egalitarian Tanzania, the figures for 1993/1994 indicated that essentially only the upper fifth of the population benefited from government spending on tertiary education.[15]

All poor together

Much of this waste can be blamed on the pervasive influence of the communist empire of a few years before. The moral advantage that the Soviet Union gained by unexpectedly launching the first satellite, Sputnik, in 1957 had been used to draw in clever students from all over the third world.* Throughout Africa I came across senior government officials in the various ministries of mines who had been trained in Russia, Czechoslovakia, East Germany and so on. The extra year that most of them spent learning Marxism-Leninism along with the local language before starting seemed a small price to pay for what was clearly the best technical education in the world.

In fact they were very badly educated. None of them, as far as I know, were given any practical experience in gold, gemstones and copper, Africa's principal mineral resources. When they saw an actual mine it would be something safely non-strategic, like coal (four mines in sub-Saharan Africa outside South Africa) and phosphates (six mines in the whole of Africa). Inevitably they were given no concept of the business of mining, nor even of its economics. This, the least plannable of activities, was to be accomplished by directives from the state. Such wealth (for had not every foreign geologist they had met spoken of the untapped mineral riches of Africa?) must not be allowed to trickle away into the hands of aliens.

The effect on the development of mining in Africa has been profound. The upper echelons of its administrators are stuffed with individuals who are ignorant of their nominal professions, yet who had been given to understand during their training that they would be in charge of determining what would be found and where. The mining laws and regulations, and the administration of

---

* It is worth remembering that the United States had announced in 1955 that it was going to launch a satellite in 1958. It was this declaration that caused the huge secret Soviet space programme to be speeded up to beat this deadline. They did so by four months. This, in retrospect, relatively modest Soviet lead in the space race led to much hand-wringing over the quality of U.S. education, to the National Defense Education Act, and to the expenditure of something like a billion dollars over the next twenty years on improving science teachers and teaching. The effect of this upon the ultimate triumph of the U.S. space programme was presumably useful, but not actually detectable.

The studious shepherds

these, have this unrealistic and unworkable concept as their foundation.

This *dirigiste* viewpoint, which is most developed, as might be expected, in the Francophone states of Africa, saw private capital as the junior partner and ignored the financial hazards —of the ultimate size of the resource, of the price the product will fetch and so on—that are inherent in mining investment. Even more alien to these administrators is the idea of a market place in mineral titles, which is how such risks are spread and minimized in countries with a healthy mining industry. After all, speculators were shot in the Soviet Union.

The failure of that particular educational system is now clear. Here was a people whose literacy rate was nearly 100%, turning out many more engineers than the United States, yet all this cleverness was wasted because that education was isolated from finance, economics and foreign collaboration. The education provided turned out to be no use outside an environment that allowed knowledge to flourish only in things the state wanted—heavy industry, nuclear power, man in space and military-related technology.

Like the Soviet Union, Africa has shown that it is possible for wealth to be poured into education, just as silver from South America poured into Spain in the sixteenth century, and with as little effect. Education, like wealth, which is acquired in a vacuum where it cannot be applied constructively, is worse than useless. Back in 1982, Buckreef's comprehensive failure was one of the many casualties.

The one consolation is that the near-unique uncoupling of the link between education and development that has occurred in Africa has been taken on board by most of the development agencies. In 1995, two World Bank members wrote—

> One of the tragedies of development is the rapid expansion of schooling in sub-Saharan Africa—at substantial costs to households and governments—that brought, on average, almost no increase in labour productivity and wages. Investing in people pays off only in an environment of expanding opportunities.[16]

Paraphrasing, it confirmed what I had found: education can be a waste of money if there is nothing constructive that the individual can do with it. For me, at Buckreef in May 1982, the demolition of the conventional wisdom about development assistance had begun. It turned out that I was not alone.

Chapter Four

'LOOK ON IT AS A CHALLENGE'

> *The history of all African effort, hitherto, is a history of wasted European life, squandered European money and blighted European hope.*
> —Charles Dicken's letter to his sponsor and friend, Angela Burdett-Coutts, in 1857.

The emotions Charles Dickens felt are now so widespread that they have acquired the dignity of a label, 'Afro-pessimism'. This is, however, used as a term of opprobrium by that fractious virtual entity that is misleadingly labelled the 'donor community'. For to use it might send the wrong signal to what might be called, with rather greater accuracy, the taxpaying community, who need to be reassured that aid is achieving something. So the first bunch regard Afro-pessimism rather as the Victorians regarded syphilis; not a subject for polite company. I am here to come out in public and state that I have a galloping case of it, but that this admission has been delayed until both the nature of the disease, and hence its cure, became apparent.

As has been explained, this cynicism was a late arrival with me; others, even amongst those who might not be expected to grasp how endemic it is, knew all about it. A majority of those who sat or sprawled that forgotten afternoon during our intelligence briefing in the then Rhodesian bush, were life-long subscribers to Afro-pessimism—the analogy does not extend to calling them sufferers from it. They were very robust Afro-pessimists.

Those farmers and shopkeepers and miners and insurance agents, temporary soldiers, embraced the creed completely. Amongst them were a few bad men, bullies, thieves and torturers, as well as numerous gentlemen, much as with the other side. But unlike the other side they all, good and bad, believed that a black government would destroy the integrity of their country and impoverish its inhabitants. Unlike the other side they were right.

All poor together

Before I went to Buckreef I had already returned to two other countries I had known before Ian Smith turned Rhodesia into a laager. Both Mozambique and Zambia were true 'front-line states', bordering Rhodesia and seriously dependent on her transit business, exports, markets and infrastructure. So once they began to provide bases for the guerrillas they were subject to economic harassment as well as the occasional raid on the camps from their neighbour.

In retrospect, nothing very much was destroyed; a dam and some bridges blown up, some camps razed. The real damage was self-inflicted. Mozambique began the process of creating a sub-tropical Stalinist state and Zambia adopted something called Humanism, dreamt up by their leader, Kenneth Kaunda. Neither system allowed for any significant economic activity outside government control, and that control was incompetent, dilatory and increasingly corrupt.

But then, as they pointed out to the aid donors, they were in a state of conflict with their racist white-ruled neighbours, South African included. Actually South African Airways never ceased to fly into Maputo and South African goods were never off the shelves in Zambia, but there were always good excuses for that.

The Portuguese had been in Mozambique since about the time that Columbus first arrived in the Caribbean. Indeed, not long after that they had managed to get some Dominican missionaries into the court of the Munhumutapa in present-day Zimbabwe. He was heir, as we shall see, to the extensive gold-based realm that had previously been centred at the stone city now called Great Zimbabwe. Until then the gold had been sold to 'Moors'; Swahili-speaking Muslims from the coast. The Munhumutapa had no illusions:

> In all the regions, or the greatest part thereof, are many Mines of Gold ... It is pain of death for any Moore which discovers a mine to take away any, besides his goods forfeited to the King ... This severitie is used to keep the Mines from the knowledge of the Portugals, lest covetous desire thereof might cause them to take away their country.

'Look on it as a challenge'

This was written by João dos Santos, one of the priests who survived the inevitable falling-out, in his 1609 book *Ethiopia Oriental*.

After the missionaries were murdered, and because the route up from the coast entailed being struck down with malaria as well as by hostile tribes, the whole effort languished for hundreds of years. For travellers Mozambique became simply a hot, scruffy and unhealthy place to get through on the way to the healthier, wealthier interior. Much of it was the haunt of slave traders. The Yao of the northern end of the country were the big middlemen in this business, and slavery was not abolished in Mozambique until 1878, just twelve years before Rhodes' pioneer column pushed up into Zimbabwe. However, as the majority were exported to Brazil and Arabia, they didn't get the publicity—or the vocal and savvy descendants—of those who survived the middle passage to the United States. (Which led ultimately to the possibly apocryphal bumper sticker that said 'If we'd known you were going to be this much trouble we'd have picked the damn cotton ourselves.')

Nonetheless, when Rhodes' pioneers appeared on the highveld (some visiting the ruins of Great Zimbabwe on the way), Portugal remembered that it had some claim to pre-emptive rights over the area. Thereupon occurred a couple of confrontations in the mountainous (and, of course, unsurveyed) eastern border with Mozambique, in the course of which the Mozambican levies were compre-hensively repelled.

One group of enthusiastic victors then set off to capture the port of Beira for the putative Rhodesia. They had nearly got there, fortress after fortress giving up without firing a shot, when a frantic message came from Fort Salisbury ('spare neither man nor horse') telling them to come back immediately. Queen Victoria's favourite nephew, King Carlos of Portugal, who was in a huff over an earlier setback in what is now southern Malawi, and had turned down a compensatory offer of the Order of the Garter, had complained bitterly to Auntie about this new provocation. But it is a pity about the port.

So Mozambique went unravished by the British imperialists and had another eighty-five years of Portuguese indolence and

All poor together

indifference. Much of the country was farmed out to semi-private plantation companies who provided just the type of exploitive capitalism to ensure that any black government there would embrace socialism with eager arms. On top of that, such black advancement that had occurred was put into reverse in the 1930s, when attempts were made to resettle peasants from Portugal. To accommodate them blacks were removed from any responsible positions—such as locomotive driving—that they had attained. In Mozambique the results of this white settlement were undramatic, although a similar scheme in Angola, where there are Zimbabwe-like fertile uplands in the south, prospered.

In 1980 I had not been back to Mozambique for thirteen years, although call-ups had taken me to its borders often enough. The Portuguese left in a hurry and what they hadn't taken they had broken. The wretched country then had seven years of incompetent Stalinism and a vile anarchy in the bush from the activities of the guerrilla Renamo movement. It was not a happy place.

I knew that, and so was prepared for the poverty, the queues and the unpainted appearance of everything after thirteen years of mismanagement and war. Nonethe-less it was an eerie trip. There was no industrial activity to speak of; no smoke came from factory chimneys, and the people slouched listlessly around the streets. The only active individuals I saw were in a mob (it could not be called a queue) besieging a bakery, trying to get at the loaves being sold there. The only commercial activity was at the ports; in Beira where coking coal was being shipped to East Germany, and in Maputo, where an ancient, fabulously ramshackle railway wagon tippler was being used to hoist and invert loads of South African coal into the bunkers of a ship alongside.

One of the bravest men I ever saw worked here. Because the original wagon braking system had long worn out, he squatted in a gully beside the hoisting deck, stopping the oncoming trucks with their 40-tonne loads of coal at exactly the right place by thrusting a metal chock under the still-turning wheels. So old was the tippler that the motors were wound for direct current electricity. The switch gear probably once had capacitors to reduce flashes, but they had gone long ago. Now the operator stood outside the

'Look on it as a challenge'

control room and poked a broomstick through a window at the switches to avoid being electrocuted; the sparks from them lit up the little room like an arc welder.

The expatriate life in Maputo, however, was not too bad. The beer was good (East German aid) and the beach that stretched northwards from the city was populated by hundreds of them at the weekend, the various nationalities each having a unwritten agreement to occupy the same section. So the three-kilometre stroll to the grubby restaurant at the far end led through perhaps thirty successive groups of French, Russians, Cubans, Roumanians, British and so on, all playing volleyball or football, splashing in the muddy, tepid water and towelling down small children.

Shortly before that I had been back to Zambia, again to do with coal. As part of that country's striving for self-sufficiency from its white-ruled neighbours, it had been persuaded to put up a nitrogenous fertilizer plant using local coal as a feedstock. The plant was not running (come to think of it, I have visited it three times now, and it was always shut down). However, it was clear that the Japanese, whose aid built the thing, had left nothing to chance. The foundations were designed to resist major earthquakes and the massive warehouses had roofs that had been designed to carry up to two metres of snow.

Zambia is famously described as having been born with a copper spoon in its mouth, thanks to the enormous copper deposits there. At independence in 1964 it had one of the highest per capita incomes in, as the phrase was then, Black Africa. Not to bore with details, the mines were nationalized (but paid for) in 1970. With nice precision, copper output peaked at 720,000 tonnes in 1969 and then started to fall; by the turn of the century it was down to under 300,000 tonnes. The grisly details can be found elsewhere.[17]

I had worked there until 1967, and the thirteen intervening years had not been kind. The details can be imagined: the crumbling roads, the empty shops, the dry, overgrown public parks. But what struck me most was the evidence of increased crime. In the leafier suburbs of Lusaka, where it had been unusual to see even a policeman, whole companies of private security guards paraded every evening, before dispersing to the homes they were paid to

All poor together

watch. These, previously lightly fenced or hedged, were now cowering behind high, raw walls of concrete blocks topped with broken glass.*

Of course, much of this could be blamed on the side effects of sanctions against Rhodesia, which cut Zambia off from its natural export routes to the south. In addition, the price of copper had been slowly falling over the years. But Zambia failed the real test, which was how well she did when these connections were in place again. The deterioration continued, from 600,000 tonnes in 1980 to half that in 1998. Its place amongst the world's copper producers fell from 4th to 7th between 1980 and 1997. The only consolation was that its 50% decline was as nothing next to that from Zaire, now the Democratic Republic of the Congo, whose output fell from nearly 440,000 tonnes to under 40,000 tonnes in that period. (The winner was Chile, which brought its output up from 910,000 tonnes to 3.4 million tonnes over the same years.)

In any event I was fairly hardened when I stepped into an Air Tanzania 737 for the first time. And I was prepared to make allowances. But it was that first return to Tanzania—a country that had not been through a war, but had adopted the same one-party socialism as Mozambique—that marked the real start of my own disillusionment.

To go back a bit, in 1958 I had been one of the few whites at the new local university, meeting and competing with blacks on equal terms, an unusual experience for us young white Rhodesians. And I found that not only was I dealing with people who were, on the whole, cleverer than I was, but that they had the usual mix of nice and nasty—perhaps nicer on the whole—as my white contemporaries.

Even my wanderings up through Africa in the early 1960s did

---

* A novelty then, but now the typical Zimbabwean bourgeois family, black or white, cowers in turn behind high walls, with the added protection of house alarm systems connected to 'armed reaction units', personal 'panic buttons' and electronically-controlled gates. Many have got rid of their private security guards; they tend to sleep through burglaries. Electric fences have replaced broken glass.

'Look on it as a Challenge'

not shake my confidence. Unknown to me I was travelling in the last years of the civil society imposed by the colonialists, and there was no hint of the chaos to come. Tanzania—Tanganyika then—was a country I liked; the people friendly and straightforward, Swahili a much easier language than Chishona, the atmosphere optimistic. A poor country, yes, but getting richer and with the glorious opportunities of independence only a few months away.

But by May 1982 Tanzania was a very poor country indeed, and getting poorer. On that first visit I had managed to forget my toothbrush, and asked Hekki Sorensson as we drove into Dar-es-Salaam if we could stop at a chemist's shop and buy one.

He laughed. 'I don't think I have seen a toothbrush for sale anywhere in Tanzania. Never mind, we brought in a container-load of supplies when we came—our government gave us a long list of things to bring. I am sure we can find you one.'

Now this was not a case of a very poor country that has never had toothbrushes still not having them. In the same city at the brink of independence in 1963 I stocked up with supplies for the next leg of my up-Africa journey, including, as it happened, a new toothbrush. This I had bought in a chemist's shop opposite the war memorial statue of the Askari; both the shop and the memorial are still there. But by 1982 there was next to nothing on the shelves of that shop, despite more than 40% of the Tanzanian government budget, and goodness knows what else, being paid for by aid money.

It got worse. The *Mwalimu* decided to give up part of his power to a new post, that of Prime Minister, the better to implement his brand of socialism. His first appointee was one of the party faithful, John Sokoine, who was to out-Herod Herod.

'The Crackdown' (as it was called) that followed in 1983 was directed at the usual suspects—that is, what the government-controlled press called 'saboteurs and profiteers'—and depended on denunciation. The scores that were settled as a result led to a wave of fear sweeping the country, and anybody with assets hastened to get rid of them. Money was thrown into the sea, television sets into the bush. The *Daily News* and *Uhuru* featured whole pages of photographs of the denounced, to the glee of the

All poor together

poor. The tall poppies all bowed their heads to avoid getting slashed. Any idea (if there had ever been one) that aid-dependency was not a normal state of affairs and that private enterprise was, had been safely buried for another five years or so. The rhetoric that the *Mwalimu* always used, about self-sufficiency being the key, remained in its place, of course.

Luckily—there is no other way to describe it—John Sokoine was killed when his car was in a collision with one driven by a military trainee from an African National Congress training camp in the centre of the country. His successor, a Zanzibari called Ali Mwinyi, called a halt to the witch hunt.

But the real damage had already been done, over the eighteen years of independence prior to The Crackdown. This was the destruction of what, for lack of anything better, might be called the self-respect of the average Tanzanian. A people who had looked you straight in the eye and who were certainly no more dishonest than the average, had become shifty and crooked. Not because they wanted to, but because they had to. Survival depended on having a second job (which was illegal), filching items from employers, standing in queues and getting as close as possible to the truly, safely, wealthy—the thousands of expatriates that the country now evidently couldn't do without.

I knew all about the corruptions of capitalism in Africa, the drift into arrogance and callousness towards employees who were cheap and readily replaceable. What I hadn't considered was the possibility that state policies could destroy the rectitude of a whole citizenry. What the *Mwalimu* had done to his people was worse than the most exploitative application of capitalism could have achieved; he had created a nation of cheats.

I was to find something similar, albeit not so dramatic, had occurred in all the countries in which I was to work over the next eighteen years. From Burkina Faso to Zambia, the ideals of the fifties and sixties—of government as the sole arbiter of almost every national activity—had become entrenched, with terrible results for the health and prosperity of the average African.

Counting it up, I had worked in eighteen African countries, usually deep in the bush. And they were all the same. Authoritarian,

'Look on it as a challenge'

if not actually tyrannical, governments, populations that were growing faster than their economies, a widening gap between rich urban elites and poor rural peasants, that was fuelling the corruption and moral decay that every country was weighted with.

Yet all these countries were still receiving ever-fresh crops of idealistic men and women from the donor agencies, teaching in bush schools, giving advice to farmers and assisting in government departments, to absolutely no avail. They were always new faces because their predecessors had left, exhausted, defeated, saying (but seldom publicly; they had careers to make in the public service elsewhere) that development assistance in sub-Saharan Africa was not working. The only advice that they gave to their successors was to look on the experiences awaiting them as a challenge.

Just how much it is not working comes from the consolidated figures for the last 50 years. Since 1950 approximately US$300 billion (in historical dollars) has been spent on the region, and at any one time since about 1965 there have been about 100,000 expatriates working there. In that period the per capita GDP has remained at about U.S.$300 per annum in constant dollar terms, probably dropping by about 5% in the latter part.

For me, ever since that first shock, the steadily nagging question has been: 'Why this awful failure, not just of showing people how to do things but of getting people to understand why they should be done?' Of course, I was not the only one asking this. Right from the early days of development assistance—in the optimistic 1960s—there were reservations. The gloomy predictions of ex-colonial administrators could be dismissed as sour grapes, but there had already been some object lessons in the difficulty of transferring technology, of which the Tanganyika groundnut scheme (Tanzania again) was the most notorious example.

It is a feature of complex and refractory problems involving human society that it is possible to make almost any Good Idea appear to work, given enough enthusiasm and a catchy name. Think of the management fads—management by objectives, theory X and theory Y, transactional analysis, strategic planning and so on. Or educational ones—Montessori, Pestalozzi, eurythmics, progressive. Or health even—remember biofeedback, acupuncture and holistic medicine? Thus it is with development assistance; an idea, or at least a buzzword, is born, followed by a flood of enthusiasm which slowly ebbs.

Of course, there was no clear jump from one concept to the next; like management theories these fads often co-exist, because the momentum of a Good Idea is sometimes such that it takes decades to become unfashionable. Further, because there is a such a lot of money seeking Good Ideas, whole institutions are often created dedicated to them, with the consequent determined resistance of their occupants to what is curiously called *restructuring*. For instance, *capacity building* has led to the Africa Capacity Building Foundation, with experts in this matter in downtown offices in Harare, and more experts at the Institute of Capacity Building in Africa in Addis Ababa. You can't just nuke them. Can you?

To go back, the first big catch phrase in the history of development assistance was *technology transfer*. This was a rational, obvious objective, and it has never really gone out of favour, for the good reason that it has the attraction of killing two political birds with one stone. Not only is the desire to be a benefactor satisfied, but the technology to be transferred involves the supply of goods and equipment from the donor country, conveniently recycling the aid money. As a result Africa is littered with steel mills, glass factories, paper mills, mines and irrigation schemes, all mouldering in the bush, often still with a dwindling complement of workers, languishing their lives away on tiny salaries.

It was deduced that a major reason for these failures was that there were none of the support services needed to sustain such technology. Buckreef Mine was five days' hard journey for a lorry by road from Dar-es-Salaam and almost everything, down to gloves and boots for the workers, had to come by sea through there. So *infrastructure* became the hot word, and roads were rebuilt, rail tracks relaid and airports modernized.

When this also failed to improve things, it was clear that hardware alone was not the answer. There was something wrong

'Look on it as a challenge'

with the system. The nature of aid meant that it went from governments (or quasi-governments, like the UN and the World Bank) to governments. Was the problem that governments were not the best operators of steel mills, glass works, etc? It seemed unlikely, or at least inadmissible. But as the evidence came in following the collapse of the eastern *bloc* that governments were indeed incompetent businessmen, views began to change.

So, *privatization* was to be the key. Unfortunately, this has not had a fair trial—few wanted to buy the decaying and debt-enmeshed white elephants. More, many governments did not really want to sell them; their senior staff were often party faithful, related to powerful ministers. In addition their employers were voters. These rightly suspected that most of them would be out of a job if privatization happened, and bent all their efforts to ensuring it did not. So at the end of this route a detour was made to a cul-de-sac that stopped at a sign marked *commercialization*. This simply meant that the parastatal responsible for the steel mill still ran it as a government body, but now it was supposed to make money. As before, actually. Thus *privatization* had limited success.

Then came *sustainability*. This had its origins in a 1987 United Nations environmental report, produced by a committee of the internationally good and great (Norway's then prime minister, Gro Bruntland, chaired it). But this buzzword has never really caught on either, although it is popular with donor governments because it suggests to the taxpaying community that their contributions are not open-ended. But since so far almost nothing the aid agencies have done in Africa has turned out to be sustainable (except with the help of further dollops of money) it is treated a bit coolly by the donor community. If the agencies said unequivocally that they were only going to fund sustainable projects, somebody who mattered might hold them to it. And what then, eh?

So something else was needed. People were the key, it was declared. Yes, of course, training, management development ... *capacity building*! The very thing. (Purists will nag, 'What does it mean?' No, no, no. It is what it *doesn't* mean that counts. It doesn't mean competence building, because that would imply incompetence. The same with ability building. And capability.

69

All poor together

Got it? And you think a bureaucrat's life is easy?)

The demise of *capacity building* will occur in the next five years; its successor is already in place. This new big Good Idea has all the signs of having been thought up by a United Nations conference of experts, its seed implanted at the beginning by one of their paymasters.

With only very little experience of this sort of thing, the scenario can easily be envisaged: the hushed auditorium; the serried, multicultural, multigender ranks of development experts; the interpreters mouthing silently in their cubicles; the television camera lights (the UN will be making its own publicity video of the event).

The hush intensifies. The bigwig opening it, a deputy undersecretary general, mounts the podium. Small, dark, dapper ... he has a message from the Secretary General, no less! A stir of excitement, but quickly, decently, subdued. A piece of paper is held up (it could be, perhaps is, a dry cleaning receipt for his natty suit). The S.G., as we Resource Persons say, has asked the gathering to (hold collective breath) consider and pronounce on 'what exactly is it that we trying to do?'

There is a gasp of relief followed by a swelling mutter of admiration. A truly profound thinker, the S.G. The conference has a theme. There will be an abundance of opportunities for earnest discussions, interventions, contention, tense gatherings in hallways and, through all this, like the makers of exquisite watches, an elite group labours over the precise wording and punctuation of the recommendations and the final communiqué. A week's work at least. Their fees are assured.

So, from this great swirl of hot air (I am still informedly guessing) came a final suite of recommendations which, with the preamble and appendices, adds up to what? twenty? thirty? pages. But the key phrase, direct, pithy and meaning nothing is in there somewhere and everyone is aware which one it is: *poverty alleviation*.

You now know the answers to the questions.

No, the phrase is not 'poverty elimination'; that ranks with 'sustainable' in its striving for the unattainable, and besides the

'Look on it as a challenge'

United Nations has had some embarrassing experiences with slogans like 'Health for all by the year 2000', which we will consider later. It is certainly not 'wealth creation'; making people rich is not the business of development assistance (although that is what it has mainly achieved amongst the elite of the recipient countries). *Poverty alleviation*; a splendid new buzzword.

Unfortunately, I—far out on the furthest reaches of the development assistance industry (good God, mining?)—was ignorant of this when I spoke in Washington in 1995 at a World Bank Conference on small-scale mining. So I was nonplussed when, after I had explained—nay, praised—the get-rich objectives that lie behind all mining, not just the small-scale sort, I was asked by an indignant delegate just what would this do for *poverty alleviation*. It seemed a silly question, but I gave the obvious answer; it would make some entrepreneurial people wealthy—the role model thing to encourage others—and that money would not so much trickle down as churn around a bit. People would get richer for a time; at least that was what had happened elsewhere.

'But the issue of *poverty alleviation*?' persisted the intervener.

'Well, it will put some money into the rural areas.'

'But *poverty alleviation*?' He was getting desperate. Hadn't I learnt my lines?

'I can't see small-scale mining seriously alleviating poverty anywhere quickly.'

There was a gasp in the auditorium. This lunatic was cutting not only his own throat but also that of number of his listeners. If small-scale mining did not achieve *poverty alleviation* status then funding for it could be cut off, along with all sorts of jobs and consultancies.

After my talk I was gleefully upbraided by a United Nations person—a female person as it happened.

'Well, you have really upset some people here. Telling so-and-so that small-scale mining would not help *poverty alleviation*. He was really put out. Didn't you know that the policy has been officially adopted?'

But later, in the question and answer session that followed, the World Banker who was also a speaker, and who was up there with

All poor together

me as I was grilled about my out-dated and refractory attitude, muttered discreetly, 'I think you are right. I don't believe that we can do much for poverty alleviation. That can only be achieved much later.'

The gap between what people say and what they believe is nowhere as marked as in the aid business. In certain special circumstances honesty can sometimes be encountered, and just to show that this refreshing feature still exists, here are some real-life instances.

First, a farewell speech when leaving permanently for a secure job in a country where the tap water is always potable:

'I have been here for two years and I have enjoyed myself, but I haven't achieved a single bloody thing.' That was said by a departing Englishman at the Mwanza yacht club.

This next was admitted to a fellow country person under special circumstances:

'She said that not one of our projects here is successful.' That was remarked by a Scandinavian of a lady who was in charge of a certain country's programmes. On our way in to the site the previous evening I had turned round to see them holding hands in the back of the Land Rover; *in pillow talk veritas*.

Now a remark to me by a fellow-passenger on a particularly fraught flight:

'It isn't just airlines they cannot run. I've put up cotton research stations in three different African countries now. None of the people in them are doing anything but waste time and money.' He was retiring next month.

This from one colleague to another on the same mission:

'We worked out that each gum tree that was planted in that forestry project had cost 600 dollars.' But he had enjoyed his two-year stay in the bush; it was such a pity that the project was not extended.

And so on. The awful reality has to be faced that the present system of providing developing countries in Africa with technical and financial assistance is not only not working, but it is producing a strata of cynical, covert racialists amongst those who are trying to implement it.

'Look on it as a challenge'

In any event, *poverty alleviation* is running fast through its doomed course, and the next candidate for canonization by the donor community can be discerned on the horizon.

It has come this time from the World Bank. From its point of view the failure of physical assistance in such matters as infrastructure was paralleled by the failure of financial assistance in such matters as structural adjustment. Clearly something must be missing in the mix. (Either that or phrases incorporating 'structure' have bad karma for development assistance.)

Of course, the human element! And a new, useful, word to hand. *Stakeholders*. Development must involve all stakeholders so that it is *integrated*. What we have been doing wrong is just building schools, when we need also to build the roads to bring the children to the school. We have been installing legal systems without ensuring the fair and just application of that system.

A name. Integrated Development? ID? No, poor acronym. Total Development? TD? Even worse, and there will be too many jokes about development being totalled. Comprehensive Development? CD? Acronym in use, and anyway it promises too much; failure will bring comprehensive scorn. What about Comprehensive Development Structure? CDS? Better, but as we have found, *structure* is a word to avoid. Comprehensive Development Framework? Yes! (And CDF is a good, bland abbreviation.)

So, to quote the press release:

> On January 21, 1999 the President of the World Bank, Mr James Wolfensohn, unveiled a new paradigm for the Bank's role as a development institution: the Comprehensive Development Frame-work (CDF).

'Mockery is the fume of little hearts', wrote Tennyson. Great fun, though. And, I believe, fair comment here. The respect I have for individual members of the World Bank, hard-working, committed and intelligent, yet still approachable and human, does not extend to the institution. Rather like the United Nations, its constituency is so broad that every action it takes is hedged about with emasculating qualifiers. Blandness rules. To achieve this and yet give the impression of moving forward, big, broad buzzword ideas are created.

Take CDF. Think this through for any one project—a new road, say, to the interior, perhaps to facilitate the export of coco beans and the movement of imports (and for that matter to help bring the children to the schools). Implementing the CDF policy would require ensuring that the road is maintained over the next decade or so (something that has not happened in the past), that the docks at the seaward end are also rebuilt and maintained thereafter, that the customs department there is stripped of its corrupt officers and new ones trained, that a system—a transparent system—is put in place so that overloaded vehicles using it are penalized, that street lights are put in where it passes through towns, that electricity is made available to power them, that drivers and pedestrians, children as well as adults, are instructed in the highway code, that buses are made available to bring the children to the schools … There is no end to it. But the idea sounds good.

Let us suppose that the World Bank was not an institution but a company in the General Electric group. How would it shape up under its awesome chief executive 'Neutron' Jack Welch, whose motto, you no doubt remember, was fix, sell or close?

Chapter Five

THE SONG OF A RED-EYED DOVE

> *Many persons hold the opinion that the metal industries are fortuitous and that the occupation is one of sordid toil, and altogether a kind of business requiring not so much skill as labour.*
> —Georgius Agricola, *De Re Metallica*, 1556.

Mining's bad press started early. Agricola above—of whom more a little later—also quoted from Ovid's *Metamorphoses* written in about AD 8:

> ... but men even descended into the entrails of the earth, and they dug up riches, those incentives to vice, which the earth had hidden and had removed to Stygian shades. Then destructive iron came forth, and gold, more destructive than iron; then war came forth.

Knee-jerkism of this sort lies behind much of the decline of mining in the West over the past thirty years, feeding on itself with increasingly rigorous environmental requirements. Now in many countries (England is one) the industry has ceased to be the sort of interesting thing that Uncle Jack did—and so a natural career opening for a likely lad like Uncle Jack's nephew—but is now a vile activity carried out by vile foreigners in far-off countries that are likely as not pretty vile themselves.

So with Buckreef. On one of my flights to Mwanza I sat briefly next to an elderly British lecturer in philosophy who was in Tanzania on some improbable, aid-funded mission to assist in ... Well, to assist in what? The already deeply philosophical outlook of the average Tanzanian? The creation of a new, improved African development philosophy? Classes in navel contemplation?

Whatever it was, I have to admit that I was rather dismissive of the usefulness of the fellow's profession in this foundered country. Doctors, teachers, engineers, yes. But *philosophers*?

All poor together

With suitable modesty I told him in turn that I was in mining, that I came from Zimbabwe. It is a description that I always think carries nice swaggering overtones of Richard Hannay and derring-do.

'Mining? What do you mine?'
'I don't actually mine myself. I tell others how to do it.'
'And what do they mine?'
'Well, gold, usually.'
'Gold, eh? Not a very functional metal. But lovely, don't you think …?'

We spoke no more, but I was left with the realization that for many people philosophy might appear a rather more useful career than one spent down a mine trying to prise a pretty metal out of the bowels of the earth. If that mine was Buckreef then I might just agree with them.

I have given three reasons why Buckreef was ultimately doomed. The first, the only one that we could do something about, was that the plant was badly designed. The second was more fundamental, because it was the wrong plant anyway. The third was that the system would not allow it to be put right.

But there was, if it was possible to be imagined, a yet greater obstacle to success. The Swedes who developed the mine so efficiently had put in underground drives and raises so that the ore body was sectioned up into a series of blocks. These were surveyed and sampled, so allowing the tonnage and grade in the mine to be quite accurately determined. On paper at least the mine had enough ore to keep going for many years.

Two years after my first visit I went down to the bottom of the mine and began to sample the reef exposed in the crosscuts there. Swedish aid had been approached yet again to pay for a programme of testing and design so that a new plant could be costed, if not actually built. With the help of a small gang of samplers, about forty samples, obtained by chiselling across the reef, were taken from every part of the ore body. These were crushed, split into two, and half were packed and put in a wooden trunk for air freighting to Harare for assay. Stamico had agreed to pay for this part of the work, but there was a not uncommon lengthy hiatus between declared intent and execution. So for many months the box languished in the middle of the departure area of Mwanza airport, getting underfoot and tripping people up, until the money could be found for the air freight.

In the meantime, some of the samples were checked in the primitive assay laboratory I had created at the mine, out of the broken remnants of the original elaborate equipment that had been installed there. Disconcertingly they gave very poor grades, even the samples from the contacts between the reef and the country rock. There was some suggestion in Dar-es-Salaam that such a crude laboratory could not produce accurate results anyway, which angered me, as it had been doing well enough in predicting the pathetic recoveries that the improvised flowsheet was achieving.

In parentheses I was, as you have gathered, quite proud of that lab. On my second or third trip there the mine was shut down because there was no diesel fuel for the generators. Casting about for something to do, I had a look at the existing assay laboratory. This was shut down too; it had electric furnaces, but as most of the heating bars were broken the absence of power was academic. Going back to my hazardous baggage of the first visit, this was why the lab needed nitric acid; the plan was to dissolve the samples in aqua regia, a mixture of nitric and hydrochloric acids, in place of smelting them.

To replace these furnaces I had the staff build simple charcoal-fired ones (charcoal burning is a major industry in the *miombo* woodland). But there was much else wrong there, including a critical shortage of the chemicals needed for fluxing the samples during the fire assay process.

Apparently there had been another assay laboratory in Tanzania, in Dodoma, about 800 kilometres away. It had been set up under a Dutch aid programme some years ago, and as the programme had ceased, so had the lab. But everything was still there, or anyway should be still there.

So Steven, Göran, Claus and I set out for Dodoma, a jolting two-day journey from Mwanza which we compressed into twenty hours by shift work, wrestling in turn with the wheel of the Land Rover. The road was, of course, atrocious, and on the full length

All poor together

of it we saw only one little group of maintenance workers, poking ineffectually at a drain with worn hoes. At intervals items of derelict earth-moving equipment from many lands squatted forlornly at the roadside. Police and army road blocks dotted our route, but during the day our white faces and expatriate number plates (red letters on a white ground beginning TX) meant that we were normally let through unhassled.

It was different at night, when our qualifications for unhindered travel were not immediately visible and the blocks seemed to be staffed by sleepy but suspicious amateurs. These I found out later were members of the *songo songo* militia, whose main advantage (to a paranoid administration) was their distrust of anything that originated outside their own village. Most of the places we passed through had a vaguely indicated interception point outside them, demarcated with a pole across the road—or a piece of string in one case. In order to keep the thrusting spears or gun barrels (elderly Simonov carbines mainly) away from our windows, we turned on the dome light as soon as trouble loomed, and whoever had the torch shone it on our faces.

We arrived in Dodoma at about one in the morning, weary and battered. The sole hotel—the Railway Hotel, a decaying relic of the days when the trains ran regularly—was dark and a night watchman assured us that there were no rooms. So we parked the Land Rover and spent a night in it that was almost as uncomfortable as the previous day there.

We found the next morning that there was accommodation—perhaps the *askari* had not liked to wake anybody up—and we were allocated rooms that had mosquito gauze in place of glass.

In tune with a fashion of the times, Dodoma was to be the capital city of the new Tanzania. Grandiose, unrealistic plans had been drawn up with the assistance, or connivance, of various donors. However, wisely, none of the government departments that were supposed to move there had done so. The area was a semi-desert and water was scarce, while the power shortage that was endemic in much of the country had reached extreme levels in Dodoma, such that the city was split up into seven sections that received electricity one day a week. The only significant

The song of a red-eyed dove

government departments actually in this presumptive capital were the Geological Survey and Madini—the Swahili name for minerals—whose labs, drawing offices and libraries had been there since colonial times.

Rather groggily we began our hunt for the lab equipment, or, more precisely, somebody who knew where it was and somebody who would let us take it; not necessarily, of course, the same person, but hope ever springs anew in Africa. So, with Steven as guide, we went to the Madini offices. This was marked by a yard containing the largest collection of broken-down Land Rovers I had ever seen; about forty of them. Some seemed to have been in catastrophic accidents, but most looked merely battered. The latter, it turned out, were grounded for fairly small reasons—a broken spring, a worn-out clutch—but some treasury rule prevented them from being either cannibalized into runners or sold.

The Madini staff were not concerned about this junk yard; the vehicles were all relics of aid programmes, of which there were several new ones under way, each with a fresh complement of 4 x 4s for the department. Anyway, the phrase 'but we don't have any transport' was convenient to everybody's lips if something strenuous in the bush, like sampling an old mine, was proposed.

We eventually found the lab equipment in a storeroom of the Geological Survey, where it had been lying for five years. Disappointingly, it was not all that much use, having been acquired for a geochemical survey aimed at detecting tiny amounts of gold in the soil. But there were a number of useful bits and pieces, balances and so on, along with some chemicals. Amongst these, as it happened, were several litres of nitric acid.

The Director of the Survey was in Dar-es-Salaam, and so, having located our goal in the morning of the day we arrived, we spent two more days there trying to contact him by phone to get his permission to take what we needed. He was, perhaps purposely, evasive but eventually we ran him to earth in an office somewhere there. Over a very bad line I understood him to say that we could take the items, but then again perhaps he said we could not. I said thank you, put the phone down, and told his anxiously hovering deputy that he had agreed, and that

All poor together

we would load up immediately and be on our way back.

It is of relevance that, in order to get petrol for the laden Land Rover, we were advised to go a few kilometres out of the city to a quarry owned by Stamico. They had none, nor any diesel for their own plant. These were fine new Swedish crushers, as was the generator, but they had produced no aggregate there for many months. We returned disconsolately to town where, after much searching, we were able to buy some fuel on the black market. Grinding down the uneven track from the quarry, I saw that it did have a little production after all; a solitary woman, with a baby on her back, was sitting in the bush breaking rock with a small hammer.

To return to that trunk of samples that had been barking the shins of passengers at Mwanza airport for a number of months. It can be understood that I felt a fatherly affection for that improvised laboratory at Buckreef, with its charcoal furnaces and scrounged chemicals, and irritated when there were suggestions that its assays were wrong.

Eventually, about seven months later, the trunk arrived in Harare and the samples went for an initial assay. The results were awful: Buckreef was said to have an average grade of six or seven grams of gold in a tonne, but the samples from the lower levels were only one or two grams. It was a true buck reef. Even Swedish aid, not noted for its financial discernment, declined to get further involved.

Yet the mine stumbled on. My own business with it had ended long before, after a meeting there at which I told the Stamico official responsible for Buckreef to shut it down immediately, before more money was wasted.

He accused me of sabotaging the Tanzanian economy. I said that it was he, not I, who was doing just that, and we had a stand-up shouting row in the embarrassed presence of, amongst many others, the lady who was then running Swedish aid in Tanzania. I was not embarrassed, I was right.

The cooing of the red-eyed dove, a sound that had formed the background to that military briefing on another planet years before, is said, you may remember, to sound like 'there is no gold,

The song of a red-eyed dove

I told you so ...' In fact the first phrase has to fit two notes, the second, four. Coo-coo, coo-*koo*-cuk-coo. If you try this out, you find that what the bird is really saying is 'You fool! There *is* no gold.'

There were a number of postscripts to the Buckreef saga. In late 1988 a member of the World Bank's mining staff phoned to say that although by then Buckreef had closed, the Government of Tanzania was still anxious to have the mine restarted. The Bank had given money for a brief technical audit by a British consulting company; would I take a look at their report? This, it turned out, proposed that a full technical audit, including an exploration programme, be undertaken. Out of the recommendations from this work, a development programme could be implemented. Buckreef might, even, rise from the dead. What did I think?

The sort of succinctness that this invited would have been neither seemly nor advisable, even for a report about a report about a failed project; a consultant could go belly-up that way very quickly. So the World Bank got half a dozen pages of measured comment and, as a bonus, a rough estimate of the hard currency losses that the mine had inflicted on the Tanzanian economy.

In simple terms it worked out that, for every dollar's worth of gold produced, the equivalent of about 1.1 dollars of hard currency had been consumed, most of it in the form of fuel oil for the generators. Of course, a lot more Tanzanian shillings had been used up at the same time, about the equivalent of another two dollars' worth. So Buckreef had cost roughly three times the value of the gold it had produced. Fortunately that was not very much.

However, there was some question as to whether all of the small amount of gold that was actually produced ended up in the Central Bank. The 1985 accounts of the Buckreef's parastatal parent, Stamico, were comprehensively qualified by the auditors, the reasons including the absence of evidence of a claimed 1.6 million shillings worth of gold. That would have been equal to about eighty thousand dollars at the time, or about seven kilograms of gold: several month's production.

All in all, Buckreef was an unequivocal disaster for everybody except myself, who gained consulting fees and, more importantly in the long run, a modest reputation for trouble-shooting. The only significant assets of the mine, after the expenditure of perhaps fifteen million dollars over sixteen years, were about a hundred tonnes of scrap stainless steel and a nearly new generator.

I could not and do not believe that the stainless steel was an oversight. Ordinary mild steel is universally used in cyanide plants and even the youthful expatriate 'experts' who had sat in succession in the scruffy office in Dar-es-Salaam must have known that. At some stage in the ten years that it took to build the mine, somebody had got together with the supplier and arranged for the aid money that was being used to equip the operation to be used for tanks which were twice, if not three times, the cost they would have been normally. The eastern-bloc educated Tanzanians who oversaw this transaction may, as we have seen, genuinely not have known better. But more probably it was to their advantage as well.

The generator must have involved something similar. When I first got to Buckreef there were four 0.5 megawatt Caterpillar generators, of which two only were needed to run the mine. That left one on standby and one under maintenance, a sensible arrangement. The generators were not new, but such equipment will go on and on as long as it is properly overhauled every so many thousand hours.

Yet ominously, right next to the generator house was a long, deep pit, filled with water and mosquito larvae. This, I was told, was for the foundations of a new generator, a single unit of 1.5 megawatt capacity from the reputable firm of Wartsila that was costing the Finnish taxpayer about $500,000. Perhaps it was intended that once this giant was in operation the old units would be put on standby. But it being Buckreef, and it being Tanzania, the existing Caterpillar generators were being run into the ground in anticipation of the big machine's inauguration. When, four years later, all was finally ready and the generator was started, the Cat units were worn out and the mine was totally dependent on the single Wartsila unit for power. In the event, by that stage all of Buckreef was barely operational, not just the old generators. This was due at least in part to much equipment maintenance having been shelved while the staff worked on the installation of the new machine.

Buckreef was not, as I had first imagined, a unique disaster. As I shuttled back and forth I began to realize that the Tanzanian countryside was littered with them. On the road from Mwanza to Buckreef was a plant (Dutch aid) that was supposed to make starch from cassava roots. Its start-up was delayed for five or six years (by which time much of it had to be rebuilt) because the non-food use of cassava had been banned by the regional authorities.

In Mwanza itself was a glass plant (Belgian aid), designed to use silica sand from near the shores of Lake Victoria. Unfortunately it used electric furnaces, for which no electricity was available; Mwanza itself frequently blacked out when one of the ancient generators there gave up the ghost. No glass was ever made.

On the shores of the lake was a boat yard (Dutch aid again) that was supposed to be building a mother craft for fishing co-operatives. Unfortunately the co-operatives had ceased to show interest, some even to exist, when it came to be launched, and it ended up, more usefully than most such artefacts, as a private ferry. And so on and so on.

This is old hat now. There can be very few in the development assistance business who are anything but cynical about the usefulness of aid (outside the truly humanitarian sort) in Africa. If there are, then I have not found them.

The aid worker's existence, often a feast of a couple of years in a well-paid near-sinecure followed by a famine while desperately searching for another, similar, post, is overshadowed by the knowledge that their activities (such as they are) are largely futile. Away from the front lines, such as Buckreef where the aid workers simply heartily wished to get their contracts over, aid administrators, project managers and the like were uniformly worried about their futures. Even back in the early 1980s each passing year was marked by a mournful consensus that SIDA (Swedish) or CIDA (Canadian) or ODA (British) or USAID (guess) were cutting back. In fact, overall, the amount spent in constant dollars kept growing until the early 1990s. By that time it was up to around 55 billion dollars annually, about 5 billion more than during the

All poor together

gloomy period I am speaking of, and by then Africa was getting the lion's share anyway.

What seems to have happened to give this impression, in Africa at any rate, is two things. First, there was a greater churning of projects. Failed ones were abandoned more readily and new ones more quickly adopted. Increasingly aid workers were seeing, sometimes to their distress, their contracts ending and not being renewed, while their project was left to founder.

Second, as the expected intellectual rewards of becoming an aid worker in sub-Saharan Africa failed to materialize, the financial ones have become more important. To the old hands, working in Africa now implies not so much altruism as either naïveté or desperation. As a result high flyers shun the opportunities, hearing no good of them as a career move. AIDS has to be added to the list of killer diseases lurking there. The roll call of countries where the risk of violence is relatively small is getting shorter; by the mid-1990s about a third of the region was too deep in conflict to be given anything more than humanitarian assistance.

To hazard a judgement, the staff in, for instance, the United Nations Development Programme (UNDP) is generally of poorer quality and considerably less experienced than it was when I first dealt with them in the 1980s. Then, joining the organization would crown a career, not undermine it. Yet, while I was in the UNDP offices in Maputo not so long ago, a distraught lady came bustling in to tell the official I was with that the annual all-in cost of their expatriate employees in that country had now reached $120,000 a head. Aid workers have become fewer; the doomsayers were quite right. That is because they are more expensive.

But this is racing ahead; now it is 1983, and I have still to learn more about development in Africa with a new project in Tanzania, at a place called Bulyanhulu, about fifty kilometres from Buckreef.

Here at that time there was no formal mine, just hundreds of pits in the ground. In them local miners, called *sokomotos* after the hand-forged picks they used, were digging out an unusually rich reef. Somebody had found it while excavating a latrine and a great gold rush ensued. When I arrived there about five thousand people were living off the mine, directly or indirectly. A small town had erupted, in whose market place could be bought imported desirables—radios, cosmetics, clothes, tooth-brushes even—unobtainable in Dar-es-Salaam.

The sequence at Bulyanhulu had followed what I found to be a common one throughout Africa at the time. The mining law of Tanzania—the one that had allowed a score of small gold mines to thrive in the past—had been rewritten by foreign experts, paid for in this case by the UNDP. The new version effectively dispossessed the latrine digger of his find, and allowed the government to decide who should have it. Private enterprise was anathema in Tanzania then, so Stamico was, naturally, the beneficiary, and that organization in turn entered into a joint venture with a couple of Finnish companies, who were in turn funded by Finnish aid.

I was recruited by one of these, thankfully a real mining company, Outokumpu Oy, to see what could be done about the metallurgy, which was looking tricky. The details can be skipped, because they are not part of this tale, but what eventually happened was—nothing. The drilling that was done showed Bulyanhulu to be the largest, the richest, underground gold mining prospect in Africa at a time of high gold prices. Yet even that was not enough for it to be financially attractive with the dead weight of Stamico as an equal partner. So, sensibly, the Finns allowed the project to languish, and ten years later, after some protracted legal battles, it ended up in the possession of a small, but delighted, Canadian exploration company who sold it to one of that country's biggest mining groups. The government share by that time had dropped to 15%.

The point is that the local people who mined at Bulyanhulu had no legal right to be there. The prospect now belonged to the joint venture. So, shortly after Stamico was granted it, a campaign to clear the site of them was started. It was not gentle. That strong-arm methods were employed in Tanzania I knew from a disconcerting experience at Buckreef.

The underground surveyor at that mine (surveying is a vital function in underground mines, telling miners where they are and what they have mined) had learnt his trade in the old days at

All poor together

Kilembe copper mine in Uganda. Idi Amin had shut this down through a brutish indifference to its welfare, but the surveyor was a knowledgeable man, and one of the few left who had known the Buckreef area in its glory days, when there had been some European miners about and a score of small gold mines in the region. I wanted to see if there was anything salvageable at a nearby site, and he led me into the bush about five kilometres away, where a few scattered remnants of a mill and an overgrown vertical boiler were to be seen.

Suddenly a single shot cracked over our heads and a policeman burst out from the brush in front of us, aiming his AK47 at me. We raised our arms briskly, but he came to a halt anyway, surprised at what he had found. He was, it turned out, trying to discourage rival entrepreneurs from taking control of his mine. The policeman, on satisfying himself that we were harmless, cheerfully took us to look at it.

This mine was not the relic of a colonial-era smallworking, but a short row of pits on a new discovery nearby, down which perhaps a score of people were working. Half the gold they were winning was given to the policeman, the other half was kept by the workers. He had ambushed us because he thought we were competitors, not onlookers. Around us, other members of his unit had their own properties, and they in turn were paying part of their take to the inspector in charge. No doubt the latter was making a contribution to the welfare of his boss, and so on up.

Accordingly, when it came to clearing the thousands of people off the Bulyanhulu prospect, events repeated themselves on a very much larger scale. Near the exploration camp was the headquarters of the police unit responsible for keeping them away. Here was a pit where malefactors were interred for, ostensibly, mining illegally. In practice the device was used *pour encourager les autres* to keep the gold coming. For the sake of appearances, however, most mining was done at night. Then the recurrent thump of explosives and the chatter of a multitude of voices carried clearly across the quiet bush to the exploration camp, disturbing our sleep in the clean, pine-smelling wooden prefabs the Finns had brought with them.

The song of a red-eyed dove

Little wonder that it was not until fifteen years after that momentous latrine was dug that the site was finally cleared, using the army. By then, however, the artisanal miners, as I had learnt to call them, had won most of the shallow gold at Bulyanhulu, and the numbers involved had dwindled considerably since its prime period in the mid-eighties.

That local people, not geologists, should find the richest gold mine in Africa may seem surprising. In fact, the record of success in locating this sort of deposit—again a hydrothermal type—through geological inference is very poor.

It was not for lack of trying. The little yellow helicopter that brought Pat Carter and Mr Vergympf to Buckreef was being used by an ostensibly Swiss-based company called Geosurvey International to look for minerals using airborne geophysics; that is, the detection of ore bodies by the effect they have on the earth's magnetic, electrical and gravitational fields. Geosurvey International's contract was with the Government of Tanzania, although its aircraft were actually based in Nairobi. However, Kenya and Tanzania were virtually at war at the time, and as the border was officially closed the Swiss subterfuge was used, fooling no one, except perhaps *Mwalimu* Nyerere.

Unhappily this sort of attempt to short-circuit the laborious business of actually sampling the ground when looking for hydrothermal reefs has worked very rarely in Africa. Indeed, as far as I know, there are no gold mines which have been found only by airborne methods. Geosurvey International detected lots of anomalies in Tanzania—places where the various fields were unusually high or low—but on drilling the likeliest ones they all turned out to be massive deposits of a mineral met with before, iron pyrites. These give a fine anomaly, and elsewhere—notably Canada—often have useful amounts of gold as well. One can guess that the largely Canadian-trained geophysicists of Geosurvey International thought they were on to a good thing. However, in Africa's greenstones, ore bodies composed mainly of pyrites are usually low in gold, unless there is a lot of quartz present as well.

Indeed, Geosurvey International and its successor, Dar Tadine Tanzania, were geologically rather naïve. That would not have

All poor together

mattered if they had been luckier, but they weren't. That curious, even bizarre, story, is told later. For the time being we are at Buckreef, and I am receiving a dozen glass slides each with a blood smear on them from the male nurse who ran the primitive clinic there.

Chapter Six

THE HEALING ART

[The goal is] '... the attainment of all peoples of the world by the year 2000 of a level of health that will permit them to live a socially and economically productive life.'
—Declaration of the Conference
'Health for All by the Year 2000',
Almaty (then Alma Ata), *Kazakhstan, 1978.*

Buckreef extracted a human price. Apart from a few accidents and a murder (the above-mentioned cyanide in a Tanzanian engineer's mid-morning *chai*), these were more medical and social than industrial. The social, or rather psychological, damage must be left to last; the mine's failure probably dealt serious and long-lasting harm to parts of the Tanzanian psyche, but in the nature of such scars it is almost unquantifiable, and so not easy to expound on.

For the medical side, however, the problems were quite specific. The place was a hotbed of malaria; all the Swedes went down with it within three weeks of arriving and were racked with fevers at regular intervals. It was endemic amongst the work force as well, many of whom were relatively susceptible since they came from less infected areas, such as Mike Lutula's Livingstone Mountains far in the south. The mine, as I have mentioned, was built close to a *mbuga,* a seasonal marsh, and the housing of the workers and their dependants, accommodating perhaps a thousand souls all told, stretched down on to this.

The control of malaria in remote communities such as mines is not an esoteric science. It can be done quite effectively by spraying all standing water with old engine oil, keeping all grass cut very short, gauzing all windows and by providing, and insisting on the use of, mosquito nets. There was none of that at Buckreef, no knowledge of control and no desire to introduce it. Malaria was a given, and that was that. Even in the dry seasons about 10% of the work force would be off sick with it, with many more during the rains. The attendant at the little clinic simply handed out chloroquin to anybody who complained of any vaguely malarial-sounding symptoms.

As an important aside, many of these were probably due to severe gastric upsets. However, as diarrhoea evokes little of the concern that malaria does, and as chloroquin usually has a beneficial effect on stomach upsets as well as on the symptoms of malaria, the true incidence of the disease was probably over-reported. I give this as a probable reason for the absence of the protozoa in the blood slides that I took back for checking in Zimbabwe, hoping that they represented a genuine sample of the complainants. Perhaps they did not. Nonetheless, for what pretended to be a modern organization, the lack of disease control at Buckreef was shameful.

Africa's status as the home of mankind has resulted it in being riddled with mankind's diseases, and the source of most of the new ones. This shows up in the statistics. The demographers use a unit called a DALY, standing for Disability-Adjusted Life Year, which is the years of potential healthy life lost due to premature death and disability. Even in 1990, before the AIDS epidemic took its commanding grip, the DALY for people in sub-Saharan Africa, at 544 DALYs lost per 1,000 population, was getting on for twice that of the next region down, which was India at 344 DALYs per 1,000.[18]

Officially, at least, malaria is still the principle threat, killing more Africans than AIDS (although the two were suspected to be level-pegging by the end of the twentieth century). We have seen what happened with education, where a combination of a population explosion and financial stringency has meant that the per capita expenditure has been dropping. The distortions in the health sector are even more marked. In simple terms, while sub-Saharan Africa spends about 5% of GDP on education, close to the 5.5% spent by rich countries, it only spends 1.6% of GDP on health, compared with 6.9% for the rich countries.

Despite the smaller amount of money involved, all the vices of the educational system also appear in the health sector—the misallocation of public resources to favour the wealthier and

The healing art

urban population, the excessive expenditure on the 'tertiary' level of health care (the referral and speciality hospitals), the selling of supposedly free consumable items (many more opportunities in health than in education) and the use of the extended family to secure scarce health services. A survey of eight African nations (South Africa would have been a ninth, but its history has meant that the richest group makes little use of public health subsidies) shows that the poorest 20% of the population gets an average of about 9% of such subsidies, while the richest 20% get 24.5% of them.[20]

So the money spent on both education and health in Africa is largely misapplied. Not helping either are the millions lavished on defence. Tanzania is no exception; after they closed down around the time of independence, many of the larger mines in the country were turned into military bases or national service camps for the burgeoning fraction of the population under arms. Guinea, another difficult country to start mines in, spent 29% of its budget on defence, and 3% on health between 1986 and 1990, despite that period being one of relative enlightenment there following the death of the dictatorial Sekou Touré.

Yet, for very good reasons, if any part of the world needs money spent on health it is Africa. It is surprising to anybody who knows only Africa how regions with a bad reputation for health, like South America and Asia, are relatively disease-free. It is unusual to use mosquito nets, you can swim in the water without being debilitated by bilharzia, hepatitis is rare, tick-borne and other animal-vectored diseases are uncommon, while the fearsome bouts of stomach trouble suffered by all travellers in Africa seem much less prevalent. Certainly there are serious diseases that are not found in Africa—dengue fever in South East Asia and Chaga's disease in South America for two—but the impression is borne out by the statistics: the burden of communicable disease in sub-Saharan Africa is two to three times that of the rest of the undeveloped regions.[19]

Anecdotal evidence is always the most interesting. The young Peter Fleming wrote about his South American experiences in a delightful book, *Brazilian Adventure*. For eight weeks he and a

All poor together

few companions lived rough in the wilds of the Matto Grosso, making their way through it largely by wading along rivers, sleeping on sand banks (where they were being constantly bitten by mosquitoes) and eating whatever food they could shoot or otherwise lay their hands on. Only one became seriously ill with fever; the rest seemed immune to pestilence. Yet a few years before that, in East Africa during the First World War, in a bush country broadly similar to the upper regions of the Matto Grosso, hardened colonials were dying like flies of malaria, blackwater fever, sleeping sickness and amoebic dysentery. The 2nd Rhodesia Regiment was withdrawn from combat in January 1917 with only 91 men fit for duty out of the 1,038 who had passed through its ranks in the previous two years.[21]

As noted, the reason for this difference is simple enough: mankind evolved in Africa, and the diseases there evolved with him. The latter are still doing so; AIDS is only one of the most recent of a string of diseases—Lhassa, Ebola, Marburg—that have all have emerged from the forests and jungles of Central Africa. Sub-Saharan Africa is an intrinsically unhealthy place for mankind.

Health aid is certainly high. In 1990 donors sent the equivalent of $2.45 per capita to health projects there, on average about twice the amount going to the rest of the developing world's population. If South Africa is excluded, 20% of the health budgets in the region were funded by aid. It evidently has not done much good; the most recent numbers show Africa still firmly at the bottom of the health league in terms of access to safe water and infant mortality.[22]

This might be a good time to note a curious feature about development assistance. This is the way in which individual aid projects are usually trumpeted as having achieved their goals (the microeconomic aspect) while at the same time the economy of the country as a whole (macroeconomics) is going down the tubes. This phenomenon even has a name: the 'macro-micro paradox'.[23]

Once involved in the industry the causes of this contradiction can be easily seen. Aid projects work under a system known, and formalized, as 'The Project Cycle'. This assumes a series of steps

The healing art

beginning with *Identification* and finishing with *Evaluation*. The Identification part usually emanates from the host government, and the proposed project is then hawked about the donor community until it gets accepted, or at least put in a queue with a reasonable chance of getting to the front of it.

Supposing the project is for a clinic in the bush and approval is granted. The clinic is, eventually, built and stocked with enough equipment and drugs to keep it going for perhaps a year or so. Staff are recruited by the government, and the clinic is handed over (a small ceremony, little national flags on the improvised dais, speeches by the relevant ambassador and the local political *chef*, the neighbouring chiefs there in their regalia, refreshments from the embassy's entertainment budget, the shiny 4 x 4s of the officials and expatriates glinting in the sun and an attentive audience sitting packed together in the dust).

When the assessment team makes their check a few months after the clinic has been handed over, all is going well. In terms of the project cycle, it is an effective project. The borehole promised by the government may not have materialized, so water is being hauled some kilometres in drums on the back of donkey carts, but that is not the fault of the project, and anyway the evaluation team was assured by the relevant minister himself that the borehole will be drilled 'very soon'. Microeconomic success.

Of course, a few months later is not the time to check. Two or three years later should show if the project is really doing any good. That is when it will be discovered that the drugs are finished and have not been replaced, much of the equipment needs repair, the staff has not been paid for months and while perhaps the borehole has indeed been drilled, no pump has been installed. This microeconomic failure, however, does not join the list of dud aid projects but instead shows up eventually in the national statistics, four to six years later (statistics in Africa are usually out of date by the time they are issued) in terms of higher child mortality, increases in communicable disease and the absence of a correlation between health spending and health. Macroeconomic failure.

It was perhaps this macro-micro paradox that led to the misguided optimism, and even a degree of nonchalance, about

All poor together

controlling HIV/AIDS in Africa. Early predictions (by early I mean the late 1980s) had it that AIDS was a disease of poverty.[24] It was therefore a rural threat, rather than an urban one. 'Poverty', as we have seen, is a knee-jerk word in the development assistance business; a programme that does not claim to assess, address or alleviate poverty will probably not get funded.

Yet the poverty argument looked pretty convincing; after all, in the towns sexually active persons had both access to and could better afford condoms. More, all previous work had shown that urban dwellers in Africa had greater life expectancies, due to better health,[25] and anyway the great majority—something like 75%—of Africans live in the countryside where in turn the great majority are very poor.

As economic beings, rural Africans could be relied upon to respond to the threat of death from unprotected sex, coupled with the availability of free condoms. Accordingly, a programme of condom distribution in the rural areas was undertaken, amounting to 500 million of them in 1996 alone. As I travelled about Africa, I found that the bedrooms of some of the most remote and primitive 'hoteli' featured a condom or two on a shelf, bedside table or (in more upmarket accommodation) dressing table. Microeconomic success.

I have to say that most of these packets looked a little dusty. Nor was there much information around on what AIDS was and what it did. Yet back in 1982 the wives of the Buckreef Swedes were tending a fellow countryman, a sailor, who was wasting away in the squalor of the hospital at Mwanza. When I asked what he was dying of, they shrugged their shoulders. 'Everything,' they said. At that time the doctors could not be more precise than that.

Still, the condoms were being distributed, 'high-risk groups'—prostitutes, truck drivers—were targeted for warnings and money was being put in to treat other sexually transmitted diseases to reduce the number of lesions that could spread HIV/AIDS. No doubt evaluation groups gave these measures a 'successful project' ranking. Micro-economic success.

But AIDS/HIV cases continued to grow and grow. By 1990 it was becoming clear that while truck drivers and prostitutes were the main vectors, the disease was making its greatest inroads in the towns and cities. A pattern appeared showing that the middle and upper income groups—those who should have been most aware of the dangers of unprotected sex—were amongst the hardest hit. People with money in their pockets were those most at risk, not the penniless rural poor. The message did not seem to be getting across, with grim implications for the economy. Macro-economic failure.

In the early 1990s estimates of what HIV/AIDS might do to a population were based on a maximum infection of 10–20% of the total.[26] This suggested that, at Africa's high population growth rates (averaging well over 3% annually) the effect would be to bring the growth rate down to about 2%. One 1988 study postulated maximums (assuming extreme promiscuity) as high as 40% in fifty years.[27] However, assuming a 10-year average incubation period, only if a 50% infection rate was achieved would Africa's numbers decline. A study that drew on this data concluded that 'zero or negative population growth is not a likely outcome for an entire country in Africa, although it could occur in areas where infection rates are extremely high if there was no off-setting in-migration.'[28]

There was, however, one ominous note. A study at the Mama Yemo Hospital in Kinshasa on the effect of an HIV counselling programme for pregnant women found that women who were HIV positive, and who had the implications of further pregnancies for themselves and for their babies explained to them, had become pregnant again in almost the same numbers as those who were HIV negative. '... these studies are difficult to interpret, but they suggest that HIV-positive women are *unresponsive* to counselling and testing programmes designed to prevent births.'[29] The italics came from the writer of the paper.

By 1999 it was clear that, in the words of the World Bank, 'a wildfire is raging across Africa'. The Bank's 1993 estimates foresaw that as many as 12 million Africans might be HIV positive by the year 2000; the actual figure in 1999 was 22 million. In 1993 AIDS was thought likely to be responsible for 3.5% of the global burden of disease by 2000; the 'worst case scenario' had it that, if sexual behaviour did not change, it might be as high as 8% by then.[30] The actual figure in 1999 was 9%.

The more accurate the data, the worse the problem appeared. A major Zimbabwean insurance company reported that 60% of the claims they were getting were caused by AIDS, and that their actuaries had estimated that, based on the last (1990) census, 1 million people (out of about 11.5 million) had died of AIDS in 6 years.[31] In South Africa over 250,000 people were dying of AIDS annually by the end of the century, and HIV prevalence amongst women attending ante-natal clinics had risen unremittingly, climbing from 17% to 22.8% in 1998 alone. The DALY figure arising from the epidemic is now expected to push the overall figure (which, as noted above, was about 644 DALYs lost per 1,000 people in 1993) up to more than twice or even three times that. Life expectancy is falling fast, and in South Africa, whose statistics are amongst the continent's best, life expectancy is expected to fall by 20 years (from 60 to 40) by 2008. Zero population growth—even a fall in numbers—is no longer impossible.[32]

A side effect of this is that Africa will have a dramatic increase in the proportion of elderly people. This will be way over the numbers of aged that are expected to pose such a threat to the viability of the social security systems of the developed countries in the twenty-first century. There the increase is of the order of 20%; in Africa it is expected to be double this.

A key problem with monitoring HIV/AIDS was the aversion of African governments to totting up the numbers. In Zimbabwe, as in many other countries, it was not until the mid 1990s that doctors were permitted to give HIV/AIDS as a cause of death. So while the economists mistakenly saw AIDS as being bound up with poverty and the demographers were tending to understate the problem, governments were in denial.

Complicating the whole matter was the sniping by interest groups which (like some governments) saw the whole business as 'a forum for Western spokesmen to deliver lectures about the moral standing of African people, their sexual habits and practices.'[33] After all, as has been noted, malaria is still the greater killer in Africa, not AIDS; why the great excitement? These lob-

The healing art

bies gained strength from the refusal of governments to acknowledge the extent of the disease officially. In 1988, for example, Zambia had officially only 536 AIDS cases out of a population of 7 million,[34] yet ten years later a new-born Zambian child was more likely than not to die of AIDS.[35] The official data was kept misleadingly low for many years; in mid-1994 the WHO figures for the total recorded deaths from AIDS in Africa were running at only (and precisely) 331,376 people.[36] Yet by this stage the real figures must have been of the order of several millions, a number driven home by my personal experience of the tragedies all around me as I travelled. Indeed, because of the official aversion to the truth, as well as the absence of the necessary equipment in most of Africa, the WHO no longer requires a positive HIV test to confirm a death from AIDS; clinical observation is now regarded as sufficient if the patient is exhibiting two major signs of the disease, such as chronic diarrhoea and prolonged fever, plus one minor sign, such as a persistent cough or herpes.

The grim reality is that HIV/AIDS infections in Africa have gone soaring beyond the demographers' worst fears, that it is a disease that strikes the well-educated and wealthy harder than the poor, that all previous extrapolations have underestimated the rate at which it would spread, that in central, southern and eastern Africa, sexual behaviour has not been modified by the threat of an infection that leads to certain death.* Yet no one, so far, has asked the question—why should this be?

There was another, bigger, mystery that might be related to this—the population explosion itself. The development assistance business has been confidently predicting its cessation for some time now. This change, called the demographic transition, occurs when the fertility rate for women begins to drop, and in other undeveloped parts of the world it is well under way. Yet in Africa it has remained high; in the 25 years from 1965 to 1990

---

* It should be noted that the further west a country is in sub-Saharan Africa, the lower the AIDS toll. For instance, in Burkina Faso the proportion of pregnant women who were HIV positive was recorded at 10% in 1993, and was no higher in 1998. In Senegal it is around 3%, and the HIV/AIDS campaign posters are impossible to avoid.

the total fertility rate (TFR) remained steady at about 6.5.[37] Some countries, certainly, had shown reductions, all of them (Kenya, Botswana, Zimbabwe, Nigeria) places where the degree of development was above the average.

Yet many of these were falls from an improbably high figure to just a very high one, from 8.0 to 5.3 in Zimbabwe, for instance. To quote the same source:

> But the comparative lack of urbanization and education does not explain everything: data available from the World Fertility Survey (WFS) for 1978–1982 indicate that urban and rural, educated and uneducated women in sub-Saharan Africa have and want more [children] than their counterparts elsewhere.

The explanations proffered were several, but seemed, to me at least, to lack conviction. 'Traditional lineage and kinship systems, gender roles and inter-generational relations contain strong pronatalist forces ...'. 'Women's fertility is normally a strong determinant of their status.' 'Child labour is also increasingly needed to compensate for declining male labour in food crop production ...' The most honest assessment came from one of the WFS investigators, who stated that the onset of the demographic transition 'appears to be determined more by ill-understood cultural factors than by any objectively ascertainable development indicators.'[38]

Another, perhaps related, phenomenon has been the success of child immunization programmes in Africa. The improvement in immunization coverage has been described as 'remarkable';[39] it is one of the few areas where health aid works well, thanks to the enthusiastic response of the population. Indeed, at the end of the twentieth century, in that very heart of darkness, the Democratic Republic of the Congo, and in the middle of its worst civil war, a cease-fire was negotiated to enable a country-wide polio immunization programme to be undertaken. Africans value their children.[40]

To return to Buckreef in the 1980s, health and education were perhaps predictable casualties, given the situation at a failed aid project like this gold mine. But I believe the greater damage was done to Tanzanian society, by the wounding of its self-respect. Many good men, and at least one woman, the potential leaders of a Tanzanian mining revival, saw their efforts as futile and their painfully-won skills irrelevant to the obdurate problems of the mine, problems not of their making. Between the incompetence of Stamico and the inexperience of the expatriate experts, Buckreef became a graveyard of ambitions. Such self-confidence as the staff had took a savage beating, and Tanzania lost its only chance for many years of creating its own home-grown mining management. Macro-economic failure.

Chapter Seven

A DISTURBANCE IN THE ROCKS

After miles and miles of granite
Comes a little bit of schist
Where the poor Rhodesian miner
Seeks the gold the ancients missed.
— Composed by Dr B. Lightfoot, Director of the Rhodesia Geological Survey, 1933, and affixed to the wall of the library there.

Geosurvey International's unhappy experience with airborne prospecting (chapter five) serves to confirm what many geologists would perhaps rather not publicize: until very recently gold mines in Africa were usually found by people looking for them on, or in, the ground. That is, by mere prospectors. High science plays little part. Indeed, they may be even more often found by accident; the discovery while digging a latrine at Bulyanhulu is a typical example and there have been numerous cases of farmers ploughing up gold nuggets from hidden reefs underlying their fields.

Whatever the circumstances, it is the local people who have found almost all of Africa's gold mines. Of the four thousand mines in Zimbabwe only a handful were found by geological inference; the rest, like my father's smallworking, were discovered centuries before the white man came. The type of frenzied artisanal mining activity that is going on now in a score of African countries was in full swing in Southern Africa when the Portuguese rounded the Cape of Good Hope; that is, about the same time that Columbus spotted Haiti. Great Zimbabwe then was the hub of a gold-mining and trading empire, just like the west African ones of Mali, Ghana and Asante* two or three hundred years before.

* Neither the Ghana nor the Mali gold empires correspond to the present countries of those names. The Asante were the dominant group of the Akan peoples of today's Ghana.

A disturbance in the rocks

West Africa came first probably because the relentless rainfall there gave a very shallow water table. Over a period of many millions of years, part of the gold (hydrothermal again) dissolved and re-precipitated in the zone around this level. This process—re-mobilization—can lead to big, irregular nuggets at the surface, clearly marking where the reefs underneath are hiding. Another reason for the early start was that West African reefs are usually richer than those of either the south or east. One theory has it that these weak zones in the old rocks received not one but several dousings of the hot, salty, silica-rich, gold-bearing water from the cooling granites nearby.

There is a famous story in west Africa, perhaps exaggerated, about Mansa Moussa, the ruler of Mali in the fourteenth century, who dispensed over ten tonnes of gold on a pilgrimage to Mecca, the gold being transported by 500 slaves each carrying a baton made of the stuff. This bit of one-upmanship flattened the gold price in the region just as thoroughly as the Reserve Bank of Australia did in the middle of 1997, when it off loaded its 167 tonnes of reserves. Probably, too, the miners of Mali were as furious with their own leadership as their current Australian counterparts.

The text books say that a German geologist, in what was then Tanganyika, found the first gold there, in 1895. Given the consistent history of discoveries by local people in Africa, this seems improbable enough to merit that useful disparagement, *eurocentricity*.

Nonetheless, almost certainly Southern Africa preceded East Africa in gold finds because the greenstones around Lake Victoria are usually covered with a hard crust of laterite, an iron-based mineral, which effectively blanks off the outcrop exposures. In the south, on the other hand, geological events had already, long before, stripped off most of the more recent surfaces, to reveal what geologists call the basement—the archaean greenstone schists in this case—with their prominent white quartz intrusions.

One of these was acquired by the syndicate that my father had joined. It was called the Blanket Mine, and there have been at least a dozen mines of that name in Zimbabwe, one of them still a very big operation. Their names have a common, simple, origin, dating back to the prospectors and adventurers who arrived in the 'Pioneer Column' organized by Cecil Rhodes.

When the Europeans came into the country (invaded is still a word I find uncomfortable to use, considering that my country was largely an intermittently contested vacuum at the time, although that is certainly what it amounted to), they ran up the Company Flag at Fort Salisbury on 12 September 1890. Then, this formality behind them, they set off immediately to find gold.

Company Flag is right; it was the Union Flag with a lion in the centre, the flag of what today would be called the British South Africa Company plc. But this was a class act; the company was instituted under a Royal Charter and the great and the good sat on its board. Rhodes' creation was familiarly known as the Chartered Company, and Southern Rhodesia, as it came to be called, was run as a business until it was granted a tolerated semi-independence with its own legislature in 1923 by the British Government. Perhaps if the Chartered Company had been profitable it would have remained a business even longer, but its shareholders were never paid a dividend.

In any event, sovereignty under Her Gracious Majesty thus established, the pioneers dispersed the next day (literally; some of the more ardent left the same day) to find the gold that had been promised them.

There were almost no geologists in the Pioneer Column, as it was called, but many prospectors. Apart from them it was composed of a motley bunch of, mainly, South Africans, Australians and Americans, along with a number of products from British public schools. Such persons had just one thing in common; little respect for academic knowledge. They wanted results. What was the ... trick ... to finding gold? Around the campfires on the two-month journey northwards, the prospectors' tongues were loosened by *dop* (cheap, fierce, Cape brandy) provided by an intent audience and the truth had seeped out.

Greenstone and granite, quartz and pyrites, it was all one, they said. Just like the eastern Transvaal, where many of them had come from, Rhodesia (nobody called it such then; if the vast un-

A disturbance in the rocks

known up there had a name at all it was *Zambezia*, but most people, Rhodes included, spoke of it as 'The North') was riddled with ancient, deep pits on good gold-bearing reefs. The locals knew where they were. Just offer them a blanket; they would show you.

It was barely a secret. The first gold mines in the Transvaal, just to the south of the new country, were at a place in its north called Eersteling, which means simply The First in Afrikaans. In 1871, nearly twenty years before the Rhodesian pioneers set off to look for gold, a famous local artist, Thomas Baines, who had been with David Livingstone on some of his travels, came to Eersteling. He recorded that a particularly good-looking reef he was shown was unique because it was the only one that had not been the site of an 'ancient working'.[41]

Eersteling soon languished, although attempts to revive it have continued off and on to the present day. However, new, better discoveries were made in the Lydenburg area, amongst the mountains to the east where the central plateau of Africa falls away to the Mozambique lowlands. Places like Barberton and Pilgrim's Rest became briefly famous, and a well-educated Irishman from the Cape who turned transport rider there, bringing goods up the hazardous road from the coast, wrote a book about his experiences called *Jock of the Bushveld*. But there, too, the ancients had worked the gold; little prospecting and no geology was needed to get started. For many years the offices of one of the principal mining companies at Pilgrim's Rest had a showcase with examples of the stone hammers used by the bygone miners along with the usual mineral specimens.

The original find at Pilgrim's Rest was made by Henry (Harry) Struben in 1871, following his discovery of the presence of ancient workings there. On the basis of this, and some good pannings, he bought two farms in the area and left for Durban. His return was to be hasty and dramatic.

Eersteling was a hydrothermal situation. So, for that matter, were many mines in the mountains of the Eastern Transvaal. At Pilgrim's Rest they mined placers—free alluvial or eluvial gold, mainly from old river banks. The Witwatersrand (it is perhaps

All poor together

necessary to note that it is almost always simply called 'The Rand') was neither, a fact which, as you have seen argued, may have contributed to the ruin of Buckreef Mine in Tanzania a hundred years later. So who found the Rand?

This is not a trivial matter. Over a third of all the gold ever produced on earth has come from this arc of mines, centred on what had once been a frowning ridge of outcrops.[42] Its discovery precipitated a war. Its development created an Australia where a Paraguay may have emerged. Its wealth allowed the country's ruling elite to practise 30 years of extreme racial segregation in the face of reason, experience and universal opprobrium.

'So many legends had grown around the question which the committee had been asked to investigate and so many conflicting rumours had gained credence that the necessity for actual facts had impressed itself on the committee.' This was in South Africa in 1938, and the committee was one appointed by the standing 'Commission for the Preservation of Natural and Historical Monuments, Relics and Antiques'.[43] The question was—just who did discover the Witwatersrand?

The note of anguish in that statement arose from events two years before. Then the jubilee of the discovery of the Rand was celebrated, and embarrassingly (to quote from the report that eventually appeared) 'the claims of various individuals to be considered the real discoverers of the Rand were widely and vigorously canvassed in the press. But, from the diversity of views revealed, it was clear that, even amongst the survivors of the pioneers of 1886, there is no unanimity on the point.'

There was not even unanimity on the derivation of the name. Some said it was called the white waters ridge because the rains, when and if they came, cascaded off its face. Other pointed out that its pale quartzites themselves resembled waterfalls, and that was the source.

Accordingly a suitably distinguished committee was enrolled in February 1939 to determine the truth from amongst the antagonisms of these scratchy old men (and women). But this itself was more than usually complicated; a greenstone schist abuts the northern edge of the Rand, and this was where gold had first

A disturbance in the rocks

been discovered and mined. The Rand itself was not a simple flattish orebody, the sole relic of a gold-rich shoreline; it was a whole succession of them, one above the other, sometimes separated by ancient lava flows or by barren mud turned into shale. As a result, different people had discovered different reefs. More, in some areas the same reef was barren while in other areas it was rich. Consequently, several of the prospectors who had stumbled across it had written off the funny-looking Rand rocks (conglomerates, technically speaking) as a gold source. Where to start?

A long time back, certainly. The committee noted that the voortrekker Louis Trichardt, after whom a town is named 120 kilometres north of Eersteling, had recorded in his diary for 1836 that local natives were mining gold in the area, and forging rings from their output.

Thereafter the record becomes muddied. The early voortrekkers associated mining, particularly gold mining, with *uitlanders*, *Engelsmans* and ungodliness, and forbade digging for it. There was little or no official mention of the fell metal until the 1850s.

In 1853 the first government-endorsed prospector in the Transvaal (at the time it was called the South African Republic) was nominated by the Volksraad, the parliament of the Transvaal. He was a Pieter Marais, an Afrikaner from Cape Town who had been a forty-niner in California and who had also worked on the Bendigo diggings in Australia. His contract provided for him to be well rewarded if he did find anything mineable: £5,000, worth now perhaps £250,000. However, the Volksraad laid down the following in his 13-item contract:

> 12. But, should it happen that the said P. J. Marais makes known any information regarding the condition of the gold mines discovered, or anything in connection therewith, to any foreign power, government or private persons, whereby the peace or freedom of this Republic might be disturbed or threatened, such action shall be punished with death, without condonation.[44]

The steady paranoia that this reveals is understandable, given the context. The Transvaal was a child of the voortrekkers, who had braved great tribulations for, as Winston Churchill noted, 'the right to larrup their own kaffirs'. They had established two independent countries, the Orange Free State and the Transvaal. They were fearful of the encroaching British Empire, which had taken over first the Cape in 1806, and then Natal in 1842. The Boers were not miners; indeed at that stage they were barely farmers. Stock-raising and hunting were their principal activities.

Just a year before Pieter Marais appeared in front of the Volksraad, the Sand River Convention of 1852 succeeded in getting the British *rooineks* to acknowledge the Transvaalers' right to an independent existence and to their own government. Yet the Transvaal was to be annexed by the British in 1877, and two Anglo–Boer wars fought thereafter, the first in 1880 and the second, the big one, between 1899 and 1902. With true instinct, in the middle of the nineteenth century, the Volksraad ruled that if any gold was to be found, no foreigner could have any part of it whatsoever.

This policy had been already been effectively in place back in 1852, twelve months before Pieter Marais was signed up by the Transvaal Government. In that year a crisis was briskly averted when an elderly *Engelsman* had found gold in the area of the Rand. John Davis was a mineralogist and presumably knew what he was about. He, perhaps naïvely, showed the sample to the then President, Andries Pretorius, who told the State Treasurer to buy it. The deal done, Davis was forthwith ordered out of the country.

Now Pieter Marais, one of the *volk*, was to look for gold for the *Raad*. His approach was unexceptional, and it is part of the circumstantial evidence about who really found the Rand that the committee of 1938 should find it surprising. For he began panning his way along river beds, looking for alluvial gold.

> His experience of gold mining in California and Australia, large and varied as it had been, seems to have been a handicap rather than a help.

This is an odd statement. Whatever its origins, the gold of the Rand is now in hard rock and panning up the rivers from down-

A disturbance in the rocks

stream is how such deposits are most easily discovered—in California and Australia. A prospector would work his way along the bed of a river, panning the gravels until they began to show 'colours'—pin pricks of gold particles sticking like tiny stars across the base of the pan when the other material had been rinsed off. The fellow of romance who is depicted panning gold crouched by the side of a stream is not a miner, he is a prospector. Unless the gravel in the stream is amazingly rich he will have great trouble making a living that way; at the very least heavy shovelling and the use of a sluice box are required for a living wage.

As the prospector comes nearer the source the colours would increase in number and size, until a yellow tail of gold was left when he had washed away the lighter stones and sand. Then, perhaps a few yards further upstream, the tail would suddenly diminish, or even vanish. There, somewhere between the sample with the good tail and the sample with little or nothing, is a source of gold. That is the place to start looking around the surrounding countryside in earnest.

However, this next part of the search requires the prospector to carry more than just a gold pan. 'Loaming' is the name of the method, and it involves sampling the soil and panning it for colours. It is possible to pan without water—women in the Turkana area of Kenya and the Fetecole area of Burkina Faso are dry-winnowing alluvial gold today, and making a sort of living out of the proceeds—but this process is nothing like sensitive enough to pick up the few colours indicating a buried reef below. In addition, should a reef be found it has to be crushed before panning, requiring a heavy cast-iron pestle and mortar. Once away from alluvial gold, a prospector's mules are heavily laden with water as well as equipment.

The original Californian gold discovery had been of alluvial particles in a stream feeding a mill—Sutters Mill. Using the same technique as Pieter Marais was employing, the group of rich reefs producing it were found. They were called—and this is worth remembering because of later events in Southern Africa—the mother lode.

The loaming technique is laborious, but then the prospectors in both California and Australia had no previous excavations to show them where to look, a point the committee may have implied when it condemned Pieter Marais' conventional methods.

Even before he made the agreement with the Volksraad, he had found gold in the Juskei River, which is only a few miles from the Rand and drains from it. However, when he went back there to resume the search the results were, according to his report, disappointing. Presumably he found nothing that looked like the sort of reefs he was familiar with, and anyway, such gold as he had panned remained fine and of low grade. His report to the Volksraad was pigeon-holed, and in mid-1855 he left for Cape Town, immediately following somewhat mysterious events arising from a burglary by acquaintances.

And yet. Pieter Marais went back time and again from lodgings in the tiny town of Potchefstroom then the seat of the Transvaal government, to the Juskei River. In the process he must have crossed and recrossed the serried pebble beds that marked the hidden Rand conglomerates. In the committee's words, 'The Juskei River attracted him strangely.' He was *persona grata* with the Afrikaner establishment, and during his prospecting contract he fought alongside the young Paul Kruger against hostile tribes, becoming known for his bravery. From this distance his sudden eclipse seems peculiar.

Supposing, then, he had found mineable gold. It would not, in that small, open society, have stayed secret for long. Then it would have been touch and go whether he was rich or dead. Of course, to inflict the death penalty on this stalwart Afrikaner would have been ridiculous, but to reward him would have blown the secret anyway. Perhaps somebody thought this contradiction through, and when Pieter began to turn up evidence that there was a reef—not necessarily the Rand itself—then that somebody decided the time had come to tell him to forget all about it. Possibly, even, his silence was assured by some sort of trap followed by a blackmailing threat. A false accusation of associating with thieves? Anyway, better by far that the secret remained that way than have the country overrun by *uitlanders*.

But by 1868 it had become clear that the deep aversion by its

A disturbance in the rocks

citizens to paying taxes was bankrupting the Transvaal. In despair the President repealed the law prohibiting gold digging, and the game was on.

Eersteling was found in 1871, the Lydenburg fields shortly after, followed by the reefs of the greenstone schist (Muldersdrif) abutting the Rand in 1874. Then came 1877 and the British annexation. An Inspector of Goldfields was appointed by the new rulers. By 1880 this official reported that 'I find that wherever I go in the Transvaal that every person who holds a farm believes himself to be the possessor of great mineral wealth and if a quartz reef of any kind crosses the property he invariably pronounces it auriferous.'

Still the Rand proper remained undetected, although at least two formal prospecting parties explored the area. Meanwhile, the greenstone just to the north of it gave rise to some rich little mines, and persistent reports were coming in of more gold to the south.

The first record of a gold discovery on the Rand proper appeared in about 1883. A prospector, J. G. Bantjes (the surname is correctly pronounced, you may like to know, *Bonkies*), wrote a letter to his brother about a new find to the south of the Muldersdrif greenstone. He had found a strange-looking bit of geology which he called *banket* after a Dutch sweetmeat made of almonds in cloudy sugar rock. In fact it was a conglomerate, a generic name for rocks which have been worn down into rounded shape through being washed along water courses, then bonded together by mud or other material and then buried for sufficient time to transform the whole ensemblage back into rock again. In the case of the Rand the formation was made up of pebbles bound together by a siliceous cement originating from fine quartz sand. This was an ancient pebble beach that long burial had turned into stone.

Wrote Mynheer Bantjes:

> I mentioned to you the peculiar banket formation which I panned a few months ago; well, I took these rocks from a hole on the farm Vlakfontein.

Did he dig the hole? Or was it already there?

● ● ● ●

All poor together

We must now jump forward eighty years. At our wedding one of my wife-to-be's brothers, about fourteen at the time and seriously impecunious, gave us, as a wedding present, an old book he had found. (He was diffident about its provenance; we suspected a dustbin.) *Wonderful South Africa*[45] it was called, published in about 1937, an early coffee-table volume, full of photographs taken at a time when the self-confidence of the white inhabitants of the region was at an Edwardian peak.

In addition to the pictures, there was a section on Johannesburg and its making. The celebrations of the jubilee of the discovery had just taken place, and there was an account of the fifty-year-old events leading up to it.

Back in 1884 a team of two brothers, Harry Struben, whom we have met before, and Fred, found an apparently rich reef—the Confidence Reef—in the Muldersdrif greenstone, north of where Johannesburg sits today. You know of greenstone and quartz reefs; Buckreef was one, my father's tiny mine was one, perhaps 80% of the world's gold mines are on this geology.

It is necessary to note here that Harry Struben was by now a major character in the Transvaal mining scene. The story goes that after he had found gold at Pilgrim's Rest—not enough to merit recovery, he thought—he was sitting in the Durban Club in late 1873 when news was brought to him that hundreds of diggers had been winning alluvial gold on the farms he owned there. Harry Struben set off for Pretoria, by now the seat of the Transvaal Government, and presented the current president, President Burgers, with a draft set of mining laws, something that the Transvaal was lacking at the time. These were adopted, *faut de mieux*, and then he and the president set out for Pilgrim's Rest.

The rains had come and the Blyde River was in flood, separating the diggings from the track leading to the area. Their approach had been heralded—indeed the diggers had, it was said, threatened to assault the president if he came their way—and the entire 1,500-strong community was assembled on the opposite bank. Nothing loath, Harry Struben took off his clothes and swam across. Standing to address the throng, as he described it, 'in a state of nature' he introduced himself as their landowner, said

A disturbance in the rocks

that he wished to negotiate reasonable terms with them and would like to introduce President Burgers.

It did the trick. The presidential ox wagon was outspanned and hauled across by the diggers with a long rope, the incumbent (who could not swim) within. A digger's committee was appointed, there was a dinner with speeches, and Harry Struben arranged, eventually, to sell his property to a syndicate.

But it was this redoubtable man's brother Fred Struben who had actually discovered the Confidence Reef, which is not just adjacent to, but is abutting—indeed almost on top of—the northern limits of the auriferous conglomerates of the Witwatersrand. In that mildewed, spineless 1937 volume that joined the towels, wine glasses, toasters and the rest of our more conventional wedding gifts, Fred was quoted thus:

> I looked up against the southern range and saw that a disturbance had taken place in the rocks.

Fred Struben had also worked at Pilgrim's Rest and so knew some gold geology. He also knew that most of the mines there had been found through developing 'ancient workings'. Was the 'disturbance' a reef, a hole or, most likely, both?

In any event, Harry secured large concessions over the farms on which his brother had made his findings, concessions which included the northern edge of the Rand, and so a number of the conglomerate beds. Fred had already found these, mainly in the form of banks of water-worn pebbles that had broken free of the *banket*. However, these pebbles carried negligible or no gold; the *values*, as miners say, are found in the silica that cements the Rand conglomerate.

To Fred Struben at that time this possibility had not crossed his mind. He thought that these barren pebbles were recent alluvial beds, like Pilgrim's Rest, and so, given the greenstones further upstream, he was hoping to find gold-bearing gravels close by. However, according to a geological booklet he wrote ten years afterwards, he went back to the pebble beds a year later, in 1885, and found undecomposed conglomerate. There was some gold in it, but not much; most of the values in the Rand are in the upper layers of the conglomerate series, from a section known as the Main Reef Group.

These conglomerate layers bend sharply upward as they near the surface, emerging from the south at an angle of about 70 degrees. Fred Struben, fossicking, as he put it, along southwards towards the limits of their concession, encountered only the successive outcrops of the largely barren lower series.

Harry was, as we have seen, a prominent and trusted figure in the Transvaal's modest mining world, and Fred had already spotted the conglomerates as possible gold sources. With the full knowledge of the authorities they ran a trial batch of 25 tonnes of Bantje's *banket* through the Confidence Reef mill in January 1886. The results were disappointing, 'practically nil' recorded Bantjes.

This is the point at which matters become confused and contentious. The Strubens were part of what might be called the Transvaal establishment, and, whilst English-speaking, were being supported by the Volksraad in Pretoria and President Kruger, who was now dourly ensconced there. Their operation was a serious and, initially at least, fairly successful one. The Bantjes prospecting party was another all-Transvaaler venture that had been doggedly working its way over the Rand for several years. However, so far all they had demonstrated was that some of the conglomerates they had found (coming, unknown to them, from the largely barren lower series) occasionally carried gold, but not enough.

Various other persons were to come forward later and say that they had found payable gold in the conglomerates at dates prior to February 1886. However, if they had they did not tell anybody about it at the time. In that month, then, there wandered in a trio of bit-players, none of them prospectors, each called George, *uitlanders* all, who claimed that they literally stumbled on the Main Reef, the most valuable mineral discovery ever made. Unfortunately, their memories of what actually happened were, by the time it was thought worth recording them, vague and contradictory.

The committee's own report is therefore not very helpful either. However, it seems that something like this happened. In

A disturbance in the rocks

February 1886 two Europeans arrived at the Confidence Mine. They were George Walker, a slight man with a drooping moustache, and George Honeyball, a young carpenter from Cape Town. They were tramping to Barberton in the eastern goldfields to look for employment; George Walker had worked there before. Fred Struben set them to building a shack, which apparently fell down, and not just once, and the pair were sent on their way.

They did not go far. The widow Petronella Oosthuizen was building a house on her section of the farm Langlaagte, about eight miles away to the south east. She was George Honeyball's Aunt Nellie, or so he said, fifty-odd years later, and there they stayed. An added incentive was that a mate of George Walker's called George Harrison, who had worked on the Australian goldfields, was helping with the construction.

However, no work was permitted on the sabbath in the Transvaal. Indeed, only since 1996 have the mighty gold mines there been allowed to operate on that day. So on a Sunday George Walker was strolling about on another section of Langlaagte, with or without George Harrison, and he, or they, stumbled over an outcrop.

George Walker claimed that was in February 1886, but the committee concluded that it was shortly before the end of March, apparently on the strength of Fred Struben's say-so in an interview he gave in a book dated 1887.

In any event George Walker or perhaps George Harrison recognized it as conglomerate, panned it, and found it rich in gold. He, or they, needed money, a matter of £60, to buy the prospecting rights from the owner, and so he—for certainly Harrison was not with him—at once set off for Potchestroom and then Pretoria, to raise the money and register the claim.

But not before he (and again, perhaps they) had shown it to George Honeyball. The latter was still alive at the time of the committee (he died in 1949, aged ninety-one), and gave evidence before it. This 'picturesque old man' as the committee described him, said that after he heard about the discovery he trotted off and traced the reef on to his Auntie's section of Langlaagte. He took a piece to Struben's mill. Here, he said, Fred Struben asked

All poor together

too many questions for his liking but that his assistant Godfray Lys was more cordial and panned the rock to see what was in it.

Godfray Lys was, incidentally, a relative of the Strubens and the son of an ex-British naval officer, who had taken a famous German geologist, Carl Mauch, to look at the Rand area in the 1860s. Lieutenant Lys had used funny-looking stones—he called them pudding stones—to give a footing to the wheels of a bogged wagon. These were probably banket, and certainly the lieutenant said they were gold-bearing—two pennyweights a tonne, he estimated, not of great interest then. Mauch confirmed that it really was gold, but the ban on mining the metal still held, and they apparently said nothing more to anyone about it. Mauch, however, then went on northwards, first to Tati, in what is now Botswana, where the goldfield he located there led to the creation of the first (white man's) gold mine in Africa, and then farther north to find the Zimbabwe Ruins and the goldfields of that country. His tale is to come.

In any event, many years later Godfray Lys remembered seeing the rich conglomerate that George Honeyball had brought him, although neither he nor Fred Struben apparently followed it up. Indeed, Fred Struben was said to have been away when Honeyball brought the sample along.

However, let us put ourselves in their position. By now they had prospected their concession for over two years. The geological pattern in the south of this area, of mile after mile of strips of successive pebble beds lying above conglomerates, had become clear. That there was some gold in these formations was sure, but nothing like enough to be viable commercially. Yet now here was this young fellow Honeyball who had produced a piece of undoubted conglomerate that was far, far richer than anything they had seen before. If it had come, as he claimed, from Langlaagte, then its source was a new line of reefs a long way south of where they had been searching and apparently unmarked by surface pebbles; close, in fact, to the latitude where Bantjes and company were prospecting. To both Struben brothers and Godfray Lys the implications would be breathtaking. They would have been most unlikely to have

A disturbance in the rocks

shrugged their collective shoulders and let the matter drift.

There may have been another reason for them to move fast. The fabled Confidence Reef was a dud. Indeed, at about that time, almost certainly after the rich banket had been tested, Harry Struben wrote to the government asking for a renegotiation of his mining lease on it, as the average grade he had recovered was only three pennyweights a ton. In mining parlance it was non-pay.

This would have been a bigger blow to than it appeared. When he found it, Harry had shown rich samples of Confidence Reef quartz to the Volksraad. The government was broke and an issue of Treasury Bonds was about to be floated to meet its debts. These would have been true junk bonds; there were no resources behind them, and their appearance would have signalled the bankruptcy of the Transvaal. When the honourable members were shown the reef and heard about the bonanza it represented, they agreed to delay the bond issue.

For Harry, the Transvaal's foremost mining magnate, it would have been deeply embarrassing to have revealed that he was wrong about the Confidence Reef, but for the Transvaal's finances it would have been a disaster. George Walker's or whoever's fortuitous discovery of the Main Reef at that time must have presented a marvellous chance of removing this threat of humiliation. It may be argued that the Strubens had a very great incentive to get their hands on it.

But we are deep in a sea of conjecture, and to regain solid ground it is necessary to quote from the actual records of the Zuid-Afrikaansche Republiek.

George Walker and George Harrison acquired joint prospecting contracts on Langlaagte on Monday 12 April 1886, after which they pegged claims there covering the Main Reef Group and then sold them, for £350. They said that they had earlier secured a private deal with the owner of that section of Langlaagte to prospect there, but there is no documentary evidence of this. Claim licences were also issued to George Honeyball.

However, in March 1886, Fred Struben said he crushed a trial batch of 50 tonnes probably from the conglomerate bed which came to be called the Bird Reef, which may or may not have been a trial for the Bantjes party. This reef is normally very close to but just above the Main Reef Group. The trial gave goodish results, about eight pennyweights, Fred claimed, equivalent to over a third of an ounce a tonne. In fact by today's criteria, it was very rich; the average grade now of the mines on the Rand (as well as the small ones in Zimbabwe) is well under half this.

More concretely, there are two pieces of supporting evidence (Mrs Harry Struben's diary and Godfray Lys' letters) to show that by May they were definitely crushing rich *banket* running at over half an ounce (10 pennyweights, roughly 15 grams) a tonne. Unfortunately, the evidence does not make clear from exactly where this ore came. In any event, it was much richer than anything they had milled in quantity before and was probably from the Main Reef Group.

By then Harry too had applied for and was granted prospecting claims over likely areas and, as word spread, a veritable torrent of claim applications developed. The nature of the mining law meant that each had to be arranged with the farm owner before endorsement by the government, the owner charging a steep rent for this facility. Hence George Walker had to find £60 to get access to that section of Langlaagte where he had made the discovery.

Nonetheless, hundreds of claims held by scores of individuals or companies mushroomed on the Witwatersrand farms. To avoid being swamped by furiously impatient claim peggers, the government declared the whole Witwatersrand area a public diggings. This meant that the rights to the minerals in the remaining areas not yet pegged could be directly negotiated between miners and farmer. Johannesburg was being born; mining camps sprang up along the 50 kilometres of the Main Reef conglomerate exposure, centred on the farms Langlaagte and Vogelstruisfontein, where the Bantjes group had established their workings. By the end of 1887 there was an embryonic Chamber of Mines, and it counted 68 companies who were developing properties on the Witwatersrand.

To return to the question of who found it, the committee fi-

A disturbance in the rocks

nally made its report in February 1941, not a propitious year. It decided that there were two independent discoveries of the Rand, that is the Main Reef Group, in 1886. The first, in March was accidental, by George Walker and George Harrison. However, they were not prospectors and did not follow up on their discoveries (the fact that they pegged their find and then sold it was evidently not regarded as sufficient). The committee therefore avoided entering the arena of dispute between the Georges and sundry other claimants.

The second find, they said, was shortly after, by J. G. Bantjes. He himself acknowledged that this was 'by a mere stroke of luck', but as he promptly started to mine it, his claim to be the discoverer was seen as senior to that of the Georges. The Strubens did not discover the Main Reef Group (they never claimed to have done so) but the committee concluded that 'by their extensive prospecting and mining activities ... attracted so much attention to this area that the subsequent discovery of the Main Reef Group became inevitable ...'. In any event, history's shorthand has it that Fred Struben is regarded as the 'Father of the Witwatersrand'.

My guess is that Fred was looking for ancient workings as much as he was looking for reefs, as was J. G. Bantjes, as were the rest of the half dozen or so prospectors with gold experience who laid claim to the finding of the Rand. It is what I would have done; after all, every gold area worked by Europeans in southern Africa to that date had been marked by such pits. I think it very likely that some were found on the Muldersdrif greenstone; enough to keep people like the Strubens actively prospecting in the area. But the Main Reef was largely buried; if it had been discovered and pitted by the 'ancients', nobody, apart perhaps from J. G. Bantjes, seems to have found their workings. The proven method of prospecting failed here, and it was left to one or other of a bunch of total amateurs, the three Georges, to stumble across it by chance.

There is no mention in the committee's report of the possibility that the Rand—the *banket* reefs—had been found and mined centuries before by Africans, which is a pity, but in the context of the times, comprehensible. The Chairman of the committee was

All poor together

Senator F. S. Malan, then a senator of the South Africa Party. Likely enough he would not be much concerned with dispensing such nit-picking historical justice, even if it had occurred to him there was something incomplete about the tales they had been hearing. White men, not black men, found the goldfields, just as David Livingstone had found the Victoria Falls, and there was an end to it. In 1948 his namesake, Dr D. F. Malan, became South Africa's first Nationalist Party prime minister, and at once initiated the process of building apartheid in Wonderful South Africa.

Chapter Eight

THE ROTTEN REEF

*A mine is a hole in the ground,
with a fool at the bottom and a
liar at the top.*
—Mark Twain.

The committee had not finished. Its terms of reference were not just to enquire into who discovered the Rand's gold, but who first exploited it.

From henceforth it was to be a contest of the giants, and the Bantjes group found themselves in the centre of it. For they had now unquestionably found the Main Reef and in June 1886 had its richness confirmed by crushing a trial batch in Fred Struben's mill. For once the origin and ownership of this ore was unequivocal. The evidence was in and serious, expensive, mining could begin.

The big money in South Africa was in the diamond mines of Kimberly, where Barney Barnato, Cecil Rhodes and a half a dozen lesser millionaires reigned. The prospectors in the Bantjes syndicate, who had toiled for years over the Rand, were in no position to find the money needed to mine the great discovery themselves.

Their first contact was with a storekeeper of that town, a Mr Alexander, who took samples from Bantjes and publicly panned them next to his shop before a number of members of that familiar class, 'prominent citizens'.

The Kimberley *Diamond Fields Advertiser*, describing the event in July 1886, said:

> The 'rotten reef' (for so it is termed) has a most peculiar formation and at first sight no one would think of testing it for gold ... The quantities we saw panned at Mr Alexander's store, though small, yielded, it was said, as much as 80 ounces to the ton ...

Mr Alexander, understandably, had no trouble over the next few days in putting together a syndicate of investors to buy and

All poor together

develop the Bantjes discovery.

However, one of the observers was a Mr J. B. Robinson, one of those lesser diamond magnates, and he did not wait around, but took coach to Pretoria. With him, it was later said, was a gladstone bag containing £20,000 in gold coins. It was also said that this money was not his but had been given him by Rhodes' financier, Alfred Beit.

In any event, within the week he had concluded the lease or purchase of several farms on the Witwatersrand. To separate a Boer from his farm is normally an extremely lengthy process, and it can only be assumed that he acquired them at such an unheard of rate because he paid from the gladstone bag. His purchases included most of the Langlaagte sections, and when the mine his company built there closed in 1930 it had produced 180 tonnes of gold. He also bought several farms some kilometres to the west of the discovery, totalling in the end twenty thousand hectares and covering eleven kilometres of the Main Reef Group. The mine that developed there, Randfontein Estates, became part of the JCI Group, and was still operating a hundred years later.

He was closely followed by another of his class, Mr William Knight. Between them and their associates, they acquired the bulk of the properties that the Main Reef Group was thought to strike along. By the time Cecil Rhodes himself arrived, the central rand farms were already in the hands of rival mining groups, and his company—eventually to be called Consolidated Goldfields of South Africa—had to be content with buying up farms in the eastern and western extremities. This was a disadvantage that was to limit Rhodes' wealth from the Rand; by the time the Boer War disrupted mining, his Consolidated Goldfields was producing only one tenth of the gold coming from it.

He very nearly did not get anything at all. His advisers, almost to a man, were American mining engineers. Their gold experience had been on the Californian mother lode and in Mexico and South America—hydrothermal orebodies all. The Rand conglomerate was something new.

Gardner Williams was the leading member of this group. A Californian, like most, he had taken his degree in 1868 at Freiburg

The rotten reef

in Germany, the only mining school in the world at the time, and had worked in Arizona, Nevada, Utah and Mexico. After a year at Pilgrim's Rest he went to Kimberley, met Cecil Rhodes and ended up in charge of De Beers, Rhodes' great amalgamated diamond mining company, for eighteen years.

So here now was this apparently fabulous gold discovery on the Witwatersrand, and Williams was asked to go and have a look at it. He knew Harry Struben and in 1885 he had visited the Confidence Reef, when he had told him, correctly, that he judged the reef not to be a particularly rich one.

The principal area on offer to Rhodes—the Ferreira Mine to the east of the Langlaagte strike—had the Main Reef outcropping for hundreds of metres, about a metre wide and running at well over an ounce of gold (twenty pennyweights) to the tonne. If anything like this exists today, it has yet to be found. It became the City Deep mine, closing ninety years later having milled over 40 million tonnes of ore.

Yet Gardner Williams advised Rhodes not to buy it. Two hundred feet down, he said, the free gold—that is, the gold that could be extracted by simple shaking devices or other gravity concentration equipment—would be replaced by gold locked in pyrites. At that time it was believed that this could only be extracted by an expensive process known as chlorination, and as the grades at that depth could be expected to be only about half that of higher up, it would probably not be economic to mine the ore there.

Echoes of Buckreef. Because, of course, Gardner Williams had known Buckreef-type hydrothermal mines in America, and that was the pattern. The pyrites in the weathered quartz near the surface usually were oxidized away, leaving the gold free, although often resting a spider's web of iron hydrates. Thereafter remobilization and reprecipitation due to the seasonal changes in the water table over millions of years had both concentrated the metal and created coarser particles. Supergene or secondary enrichment is what the geologists called it. So the upper part of a hydrothermal orebody frequently has gold that is both higher grade and easier to extract than the lower levels. The original miners at Buckreef had only mined to one level, where we encountered the

All poor together

bats, because below that the grade was poorer and the values largely tied up in pyrites.[46]

Thus Gardner Williams. And he was right, but right for the wrong reasons. The Witwatersrand gold *was* difficult to extract below the first level, but not because it was in pyrites like Buckreef; there were relatively little sulphides and even less gold tied up there. No, this was a solidified alluvial deposit, and once below the secondary enrichment most of the gold particles were too small to be concentrated satisfactorily by gravity devices. The only way to recover fine gold then was by amalgamation with mercury—and the modest amount of sulphides present in the conglomerate caused mercury to 'flour', to become a finely divided solid in place of a metallic fluid, preventing the amalgam forming.

Indeed, three years after Gardner Williams gave his opinion, the Rand nearly closed down altogether. The secondary enrichment had been mined out, the ore below it floured mercury and its grade of 10 pennyweights (or less in many cases) was not enough to support the cost of underground mining and chlorination. It was the discovery of the inexpensive cyanide extraction route in 1890 that saved the Witwatersrand goldfields from abandonment.

In any event, Rhodes turned down the Ferreira and a number of other valuable properties, apparently on this argument. If the cyanide process had not appeared on the scene, he would have been acclaimed for his foresight.

In his defence it was said that he was preoccupied with the amalgamation of the four Kimberley diamond mines into one massive company under his control—de Beers. In fact his mind may also have been focused on another, even greater, possibility. The Witwatersrand beds, as the series of conglomerates were then known, seemed to be a long-buried version of the Pilgrim's Rest alluvials. Rhodes' brother Herbert had been a digger there; indeed his group, who called themselves the Pilgrims, gave their name to the town. It was starting to appear that the Rand was a vast, fossilized alluvial deposit—a palaeoplacer.

So where then was the mother lode?

Chapter Nine

RUINED BY
DRILLING ...

The vast extent and beauty of these gold fields are such that at a particular spot I stood as it were transfixed, riveted to the place, struck with amazement and wonder at the sight, and for a few moments was unable to use the hammer. Thousands of persons might here find ample room to work this extensive field without interfering with one another.
—Carl Mauch describing the 'Mashonaland Goldfields' in a letter to the Transvaal Argus, *1867.*

Carl Mauch has been described as 'a tremendous talker, who learnt what geology he knew from text books ... a mine of misinformation ... He was forever pausing in the course of his travels and declaring dramatically "Gold will be found there".'[47] Another source, more obliquely, noted that 'Records seem to show that his ambition was beyond his reach ...'[48] But nobody denigrated his determination and energy. After arriving in South African in 1865 he discovered both the Tati and Lydenburg goldfields—or, perhaps more to the point, identified the pitting of the ancients there. His words were regarded with respect; after all, he was the very best sort of mining promoter, a successful one.

• • • •

All that day the three of us had been panning our way slowly up the gravel banks lining the Pra River. Now as the evening approached we were sitting dog-tired in the clumsy wooden launch, drinking warm Club beer as its big old motor thumped it back downstream to the mouth. Richard, Charles and I had been brought together as a team to take a look over our client's concessions in southern Ghana, some years before the gold boom there of the 1990s. Since the only faintly worthwhile possibility in this area of mud and deep sediments was gold that might have been brought down the Pra from reefs further upstream, we had spent

All poor together

the day in the sun crouching by the sluggish river, panning until our arms and backs ached desperately.

Yet there is an element of satisfaction in panning, whether for gold or gemstones, and whether successful or not (usually not). Certainly the labour involved is considerable, first digging into hardened gravels, then ten or fifteen minutes of working the pan in the cool river water and then five minutes or more of careful finishing, repeated perhaps twenty times in a day.

If you are, as we were, trying to make a rough estimate of what might be in those gravels, the pan has to be filled almost completely to start with, always to the same level. Then you sink the heavy dish into the water and rake the load with your fingers, allowing the fine mud and sand to wash out in cloudy surges into the river. Eventually, after much working over, the billows dwindle away, leaving a pan full of gleaming pebbles.

Now comes the brutal business of rejecting these shining but barren little rocks back into the river, done by dipping and shaking the pan and carefully brushing them down and over the lip so that the finer particles between them all have a chance—provided they are heavy enough—to filter down between the stones to the bottom. Don't worry too much about missing a nugget; if there is one even a thousandth of the size of the stones you are rejecting it will go snuggling down safely into the base of the pan at the first shake. But there won't be.

Eventually the pebbles are all gone, and now comes the more precise business of getting the finer stones and coarse sands away. Now you are swirling the pan, leaving a long sweep of rejectable matter behind a dark tail. This is fingered carefully out, washed, swirled, and fingered out again.

Eventually all the extraneous material has been washed away and you are left with a little, or sometimes a lot, of purplish to dark sediment around the base of the rim, with a pale splayed fringe at the front, the remains of the lighter sands.

The dark colours come from the heavier minerals. These are usually composed of magnetite, an iron oxide, but often there are tiny garnets, pieces of rutile, sometimes minuscule rubies, even diamonds if you are lucky. Normally these gems have little or no value, but for the serious prospector their presence is always worth recording.

Now you can stand up and ease your aching back. If you are lucky and are near enough a reef for the gold to be coarse, then you will already have a golden tail stretching ahead of the heavy minerals, perhaps composed of a half a dozen or so flattened and distorted gold particles. But normally when prospecting the gold will be too fine to present itself so conveniently.

At this point some people use a wash bottle to spray a fine jet of water on to the concentrate, to push it back to reveal the gold particles beneath. My method is to give the pan a series of gentle impacts with the palm of my hand to cause the gold to emerge from under the heavy minerals. This motion simulates, I am pleased to fancy, the action of a shaking table, still the best gravity concentration method for fine gold.

After this comes more careful finger work to ease away the heavy minerals, and finally a gentle splash of water on the golden tail. This spreads the colours across the gleaming bottom of the pan for counting. However, the really professional prospector delicately swirls the residue and then uses a precise flick of the wrist to change the direction of the flow and spread the tiny colours over the surface, free of the heavies, all in one motion.

On the Pra that day there had been some colours, three or more to a pan sometimes, but very, very fine. Still, these gravels must be three or more metres deep. It seemed to me that there could be some good values at the bedrock, and these banks went on for miles. Why, it might end up a very big operation. All it would need was a longish backhoe to prove it.

Richard looked at me and raised his eyebrows. He was a very experienced geologist, and had some experience of the needs of clients with concessions to sell. I was new to this.

'I was in Queensland once,' he said. 'North of Townsville looking around for a client, quite a big exploration company. Found a lovely wide reef, dark quartz, vuggy, good long tails in the pan. Ounce a tonne stuff. Sent a telex to Sydney, describing it and asking them to send a drilling rig.' He shook his head.

'No?'

All poor together

'No ways. They said come back at once. So I got back and saw the boss. He was very pleased—waved my telex at me. 'Just what we want,' he said. 'Five hundred metres of strike, a metre wide, good colours in the pan. You've done well.''

'Are you going to drill it?' I asked. He looked at me, surprised like. 'Drill it?' he said. 'Drill it! You don't drill a good prospect. You sell it. There's many a good mine been ruined by drilling.'

•   •   •   •

Drilling was in its infancy when Karl Mauch was prospecting, but he probably knew that the mining industry is based on a casino that uses the perceived value of mineral rights as its stakes. The discoverer of a mine is rarely the one who develops it. Almost always there is at least one transfer of ownership, sometimes half-a-dozen or more, before the property ends up in the possession of a group with the resources to bring it into production. And each of these deals is a gamble, of course, because it is impossible to know what it is really worth until it has been mined out (although understandably many geologists, who, like all of us consultants, have especial reasons to wish to be regarded as professionally infallible, are uncomfortable when this crude fact is bruited about).

The prospecting company we were working for on the Pra was a very small one. It had acquired the concession, not the best, very cheaply in those early days of the Ghana rush, and could, perhaps just, afford our fees. If our report spoke of the presence of colours and of the possible resource underneath, then it would be well pleased, for it had something just a little more tangible than imaginary lines on a map and a licence to say it could prospect there.

It would accordingly issue a press release that read something like this:

Ruined by drilling

GARGANTUAN RESOURCES INC.

Alberta Stock Exchange Symbol GAR

Significant alluvial discovery on the Pra River Concession

A team of three mining experts, led by the distinguished geologist, Dr Richard Turner, has determined that the extensive gravel beds in our 500 square kilometre concession covering the Pra river area in Ghana are gold-bearing. These beds are believed to extend for over twenty kilometres, and at a preliminary estimate could amount to 25 million cubic metres of auriferous gravels. The Dankwa area upstream has been a major alluvial gold source over the past hundred years, during which time about 1.2 million ounces were produced ... a programme to determine the extent of the resource, and in particular the potential of the near-bedrock values, is to be implemented shortly.

By Order of the Board
Cleanheel Scuttlerun
President and Chief Executive Officer

And at the bottom will be something in smallish print along the lines of: 'The Alberta Stock Exchange has neither approved or disapproved of the information given here.'

This announcement, almost devoid of new facts, nonetheless gives a useful kick to the client's share price, enabling him to offload a few of his own shares profitably, and so perhaps, amongst other things, pay us. It will also bring other players of the game, some with rather more money, snuffling around the stock. These are developers, and they are the fellows who can afford to think seriously of bringing in a back-hoe for some preliminary pitting to the bedrock.

Unfortunately, from our point of view they will also bring in

All poor together

their own consultants. 'It's not that we don't respect your judgement,' they will say to us, or Richard in this case, by which they mean, politely, we don't think you are a crook, not yet, but ... 'But, our shareholders need to be reassured that an independent *due diligence** has been carried out.' And this new bunch they send in will be total incompetents who have assuredly never stepped outside Denver or Bristol or wherever before, let alone been to West Africa. Or so we will mutter to each other. But then such are the ways of mining promoters, and of consultants too.

Carl Mauch was, as noted, a promoter, and his 1867 letter to the newspaper quoted above may just have needed the Alberta Stock Exchange disclaimer under it. Nonetheless, twenty years later, when he was already dead from an accident in Germany, his share value soared. A very big developer indeed, Cecil John Rhodes, believed in him.

It is not uncommon that truths dealing with humanity are found to be composed of three levels. On the surface, the tritely obvious one, with a contradicting one at shallow depths. Then, deep in bedrock, is the real thing, a monumental, unyielding fact that encompasses the upper two perceptions. Not quite hypothesis, antithesis, synthesis; more the manifest, the subordinate and the reality. So, business competitors are rivals at one level, co-operating in keeping out newcomers or doing a little price-fixing on another, yet, *au fond*, clawing each other aside in the struggle for survival.

Thus it was with Cecil Rhodes and 'his North'. Gold was the obvious, the easy answer when people asked why he was so obsessed with getting there. But then there was the question of outwitting the Boers and their aspirations. Much of what he said confirmed this. Thus—

> If we get Mashonaland we shall get the balance of Africa.

Balance of power, obviously. But this is *Rhodes*, for whom power, as far as he was concerned, *was* money. Gold is money in

---

* Once a verb clause, now used as a noun to give dignity to the necessary mundane business of checking the seller's claims in detail.

the ground, and looking through his writings it is clear that about three years before he finally launched his 'Pioneer Column' into Mashonaland he caught the scent of something stupendous up there:

> Saul he went to look for donkeys, and by God he found a kingdom!
> But by God Who sent His Whisper, I had found the worth of two!

This is the tale of what today would be called a hidden agenda. I know that it is true, because in Rhodes' place, I too would have wanted to believe in Mashonaland's awesome secret, while at the same time I would have done all I could to prevent anyone else from guessing. Hence the layered truth.

Rhodes thrived on contradiction. At Oxford he took eight years to gain his degree, in part because he spent so much of his time dealing with the practical problems of the Kimberley Mines (and in part because he was bad at learning Latin and history). But on the diamond fields themselves he was all dreams, extra-ordinary ones, about extending the British Empire to all sorts of unlikely places, the Levant, for instance, and bringing the Americans and Germans into it.

One of the curious things about this very political man, this dreamer and schemer—and, when it came to the push, liar—was that those who worked with him venerated him for the rest of their lives. The memoirs and biographies of his associates, sensible fellows, bankers and lawyers and mining engineers, some written nearly forty years after his death, echo with the enduring devotion he inspired and their fury at what they regard as slanders on his character from persons of a more liberal outlook. They thought he was, literally, a great man. Unfortunately to history now he is more of an uneasy combination of Jimmy Carter and Richard Nixon. Certainly there is a huge gulf between his cynical working of the financial world and the dreams he tried to bring into reality.

'His North', as everybody now described it, beckoned. The red on the maps of the world that signified the possessions of the

British Empire were a mere lacquered toenail in southern Africa in 1886. British territory stopped at the Orange River to the east and west and on the Ramathalabane River in the centre, where Botswana (then the Bechuanaland Protectorate) bordered barbaric Matabeleland. But in the north there was Egypt (marked in stripes because it was a joint venture with the Egyptians), the Sudan (an enigma; it was still eight years before General Gordon's martyrdom in Khartoum was to be avenged by Kitchener, a friend of Rhodes') and the embryonic British East Africa that would one day be Kenya and Uganda. The thing to do would be to link it all up, all the way to Egypt. Then there would be untold, marvellous opportunities for all those good fellows he had been working with ever since he arrived on the diamond fields nearly twenty years ago, when he was eighteen. Plenty of opportunities for their millions of adventurous compatriots. And himself, of course.

There is no indication that Rhodes was acquisitive in the sense of personal possessions. He wanted money because he wanted the power to do the great things he was dreaming about. Opening up a continent was not going to be cheap. His resources, great though they were, would be far outrun by what would be needed. Where to find it?

Of course, Rhodes was now a gold bug, everybody knew that, and everybody knew that there was gold up there in Zambezia or Rhodesia or whatever it was to be called. Reef gold, it seemed to be, like the failed Eersteling and Barberton fields of the Transvaal. And the unlucky Confidence Reef. Well, let Rhodes have his dreams; he couldn't say he hadn't been warned. Let's hope for his shareholders' sakes that he is right, for it was said he was gathering together several hundred young white men to look for the stuff up there.

The evidence that Rhodes believed that Zimbabwe was actually the mother lode of the Witwatersrand, and that this was the secret mainspring behind his drive north, can only be deduced indirectly. One strong piece of negative evidence is that his closest technical advisers, Hays Hammond, Gardner Williams, J. B. Taylor and J. S. Curtis, all of whom have either written biographies or recorded their work with the great man in some detail, never mentioned the idea. Yet three of those four were Californians, who needed no prompting about the likely relationship between alluvial and primary gold. The betting is that it was a great secret in that tight circle.

The Californians no doubt remembered the enormous amounts of gold that poured forth from the mother lode area upstream of the American River, where James Marshall had first seen gold in the race of Sutters Mill. Another itinerant carpenter, as it happened. There, in one year alone, 1852, the state produced nearly 4 million ounces of gold, 120 tonnes of it, worth today over 1.3 billion dollars.

The late 1880s must have been a tense period for Rhodes, laden as he was with the knowledge of this great opportunity, yet forced to wait on the outcome of laborious negotiations with the Matabele chief, Lobengula. Mashonaland, where Mauch had been in ecstasies over the goldfields, was a vassal territory of the Matabele, and Rhodes' partner Rudd was having great trouble getting a concession for it out of 'Old Loben'.

Worse, other fortune hunters were hanging around Lobengula's great kraal at Bulawayo, trying to do the same thing. In addition, the Boers were talking about moving in to settle, if only to stop Lobengula's impis from harrying the northern Transvaal. So no whisper could be allowed to get out of the new-found reason for believing that Mauch's ravings were right, that there was indeed a vast bonanza up there.

Support for this theory comes in another 1936 book, one published specifically to celebrate the 50th anniversary of the finding of the Witwatersrand.[49] Talking about the rapid diaspora of Rhodes' pioneers from Fort Salisbury to hunt for gold, the author says:

> The experience gained in the gold rushes on the Witwatersrand and elsewhere in the Transvaal led to the idea that there must be, somewhere in Mashonaland, a main reef, or, as the Americans in the expedition termed it, a 'mother lode,' and all efforts were concentrated on the search for this.

So there it was. The critical secret that Rhodes and his advisers were hugging to their breasts did not actually require much insight from anyone who had worked on goldfields before.

However, the most convincing evidence of these private expectations of a treasure trove bigger than the Witwatersrand comes from the mining law that was brought into effect as soon as the flag was run up the pole at Fort Salisbury. It could only have been drafted by somebody who was quite certain that there would be abounding wealth for all, prospectors, developers, miners, notwithstanding the presence of pyrites, the need for deep mining and the wild remoteness of the new fields. For, if this El Dorado did not materialize, then as we will see, Rhodes' mining laws would obstruct, not encourage, mining in the new country.

Certainly there was need for a different system from that under which the Witwatersrand developed. In the Boer republics of the Orange Free State and the Transvaal, mineral rights (and the rights to the air space above, incidentally) belonged to the landowner. Any deals that were cut on the mineral rights had to be with him or her, and as the interests of farmers and miners are usually antipathetic, negotiations could become very fraught.

They have become even more so since that time. Primogeniture is not a concept that sits happily with Afrikaners. Land is customarily split up evenly between the male children and this is, or was, coupled with a habit of large families. The result has been that any big new mines in what used to be the Boer republics can usually only be started after a multitude of owners have been located and corralled in. The owners of Crocodile River, a platinum mine started in 1988 in the Transvaal, had first to negotiate with over 400 different landowners.[50]

So the new Rhodesian mining law dealt briskly with the idea of the landowner holding the mineral rights. He didn't. They belonged to Rhodes' Chartered Company. As a direct result, prospectors in Zimbabwe are still allowed to fossick on any private property without needing the permission of the land owner (although for the last decade it has to be more than 100 hectares in area for this boon to apply).

My very first introduction to prospecting, in 1949 when I would have been nine or ten, was memorable for this very reason. My parents ran a small hotel or big boarding house in Salisbury, popular as 'digs' for people trying to find their feet in what was quite a boom city at the time, and for the likes of those who had knocked about a bit and needed somewhere unpretentious for a temporary base. One of these was an elderly man (or so he seemed to me; he was grey-haired with a pocked face) called Delabilliere or, familiarly, DB. DB was a prospector, amongst a number of other professions, including that of naval officer. That career had, according to him, been brought short by a court martial held in the Great Cabin of the *Victory* at Portsmouth. Anyway, one Sunday afternoon he undertook to show my father, then a novice, what to look for when prospecting for gold. I came along for the ride.

We drove out along the Bulawayo road and turned off on to a track running along the base of a range of low well-treed hills, composed of what I now know to be banded ironstone. We parked and scrambled around the side of the biggest one until near the top we found an old adit, overgrown with vegetation.

'Let me go in and take a look,' said DB. He switched on his torch and vanished. He reappeared shortly, moving quickly and surrounded by half a dozen angry hornets. He seemed to be unmoved by their fierce stings; in his hand was a couple of pieces of ironstone veined with quartz.

'Eyes in there. Might be a leopard,' he explained succinctly. 'Let's go.' We moved briskly away down to the base of the hill to where we had left the car. DB got out his pestle and mortar and began crunching away on his sample. My father and I gathered around.

'Just what the hell do you think you are doing? You can get off my farm at once.' I looked up. A very red-faced man in a blue open-neck shirt and khaki shorts was glaring at us.

DB rose to his feet. 'My name is Delabilliere and I have a prospecting licence,' he said with dignity. 'I trust you are familiar, as a landowner, with the mining law of this country.'

'Of course I'm (something) familiar with it,' snarled the farmer. 'You can't just come in here. You prospectors are all the same. Now get off my (something) land.'

'I have twenty years of prospecting experience, and I have never been treated like this,' said DB indignantly. 'I don't believe you know what you are talking about.'

'I (something) well do. You have to ask me first. I know the law. I've been farming a great deal longer than you've been prospecting. Now are you going to get off or do I have to call my boys?'

It must have gone on for ten minutes or so, much to my awe, for I had never seen grown men really angry before, but in the end he made DB tip out his mortar and we retreated back to town. However, when we got back to the hotel DB gleefully produced another piece of reef he had put in his pocket.

'I'll get that idiot of a farmer,' he said. 'I'll peg that adit and start mining. Put a big compound for the labour there. Keep the stamps going all night, thumpety, thumpety. That will teach him to tell me my mining law.'

But when he panned the rock the tail was tiny; less than a pennyweight, DB thought.

'Means nothing, of course,' he said. 'Need a lot more samples, get some proper assays. Can't pick up fine gold like this, must amalgamate it.' But I don't think he ever went back.

Years later I heard that he had been killed by a lion in Tanganyika while he cycled through the bush. I would guess that he was prospecting somewhere near the Lupa gold region there, which was having a bit of a new gold rush in the fifties.

The point is that as far as Rhodes was concerned, both DB and the farmer were wrong. DB was supposed to tell the farmer of his intention to prospect. But the farmer could not throw him off the land if he had a prospecting licence. No permission was needed, just notification. Actually I'm not sure that DB did have a licence, but it is a long time ago now.

I sometimes drive past the scene of my inauguration into mining, but I have never scrambled back up to that adit. The population of Harare has more than doubled in the past seventeen years, so the hillside is bare, the vegetation gone for firewood, and the closely ranked houses of the townships have pressed on past it. The extensive barracks of the Presidential Guard are uncomfortably close.

•   •   •   •

There have been other confrontations. It is fifty years on and David

Ruined by drilling

May and I are standing on the closely grazed sward next to an illegal artisanal mine. Dave is a Zimbabwean gold smallworker descended in turn from smallworkers and we are in south-west Kenya, in Masailand, just twenty kilometres from the Masai Mara game reserve, near a trading centre with the Tolkienesque name of Lolgorien.

The miners are Luo, stocky black nilotics, who have flouted a ban on mining here imposed by the distant government in Nairobi on this, our mutual client's concession. In fact, not only have they been mining illegally, but they are doing so courtesy of the Masai tribesmen, whom they are paying for the privilege.

Dave, who knows the miners of old—he has spent over three years here, trying to start up a tiny operation that would have taken six months to get going in Zimbabwe—tells them to leave. The Luo protest vigorously. 'We are hungry! What else are we to do? We cry to you to help us!'

They look well enough fed; they are probably mining the equivalent of half an ounce of gold a day between a score of them, enough to earn about 700 shillings a day—say ten U.S. dollars. Even allowing for the tribute of about 200 shillings a day they pay to the Masai, that is well above the minimum wage.

Down the grassy hillside, between the scattered but dense thickets of vegetation, come the Masai; three *moran*, young warriors. Two of them are in full kit, with their hair a shiny crimson from ochre and fat, earlobes gaping, the distinctive scarlet blanket draped over their lanky forms and smelling fiercely of the cattlepen. Two have clubs in their hands, the third, disconcertingly, an old beach umbrella. They have a more forceful grievance. The Luo obligingly interpret their Swahili.

'They tell you to go otherwise you will suffer.'

Dave knows these particular Masai well; he is the veteran of many such confrontations. He told the Luo, 'Tell them that the minister and the mining commissioner are coming. He will be told about this. You are breaking the law and so are they. This'— he indicated the scattering of auriferous quartz at our feet—'belongs to the government. The government has given us a licence to mine it.'

All poor together

It was all true; we had met the mining commissioner in Nairobi to arrange the minister's visit, and minerals in Kenya belong to the state. But the Moran were underwhelmed. Their tones grow sharper 'They say you go,' said our interpreter urgently. 'You go. Or else. These people, Masai, very dangerous.'

'We will tell the minister.'

Ludicrously I had to stifle a giggle; it occurred to me that, being Africa, it was just possible that one of them, perhaps the *moran* in mufti who had some English, would have replied, 'I *am* the minister.'

But our client was with us; it would be dreadful if anything happened to The Money. We left in as marked a manner as was possible and climbed back up to the hill to the Land Rover.

Yet, unexpectedly, Dave had found that these slender, homicidal pastoralists, 'the Apache of Eastern Africa' could be turned into fine underground workers. Short of the necessary lions, whose killing established their manhood, the rigors of mining seemed possibly to become a rite of passage instead. Just as on the Witwatersrand.

• • • •

Not quite all the gold mines of Zimbabwe are based on ancient workings; there are perhaps a score that were found using geological methods, including several recent ones. However, these techniques were not necessarily as sophisticated as might be expected. Take the case of Bill West, for example, who died in 1984.

He was the thinking man's smallworker, operating a group of little gold mines on the western limits of Rhodesia's gold belts. The principal one of these was the Leopard, which sat on the edge of where the greenstone vanished under the deep sands of the Kalahari. The Kalahari desert, which extends westward from Botswana into Namibia, is, or perhaps was, a desert in retreat. As a result much of the forest and savanna woodland of a vast area, stretching southwards and westwards from Zimbabwe into Zambia and Namibia, grows on—a pleasant geological term—aeolian (i.e., wind-blown) sands. The dunes have been flattened out by aeons of rainy seasons, but their ghosts linger, visible from the air at certain times of the year as subtle differences in the vegetation, marking the ancient troughs and peaks.

Bill West was thus faced with a prospecting challenge; how to search for gold under thirty or so metres of fine sand. He had already found on the Leopard, as he followed the reef under the *gusu*, as the sand is known, that the knobthorn tree had a tap root that could extend for as much as seventy metres down to the water table. It had an unusually acid sap, and while it was unlikely to dissolve gold, it should pick up any sulphur from pyrites, and that could indicate a gold association.

So he sent out foragers to locate and bring back twigs from knobthorn trees over a vast area, ashed them and checked for sulphates. Out of the results came a pattern of values that suggested a reef structure far below. Bill drilled this, and thus found what became the Lion Mine.

Also on the Leopard Mine he had seen termites at work at the contact between the sand and the original land surface. They were also looking for water, and the grains carried up by the termites to form their mounds on the surface might therefore contain gold signposts. Bill sent out his foragers again and put together a washing device that allowed samples from eighty termite mounds to be simultaneously checked for gold. Eventually, after years of work, came another mine, called, of course, the Termite.

To get back to Rhodes' scheme. At first sight the mining law he had instituted was a dream from the prospector's viewpoint. The catch was in what aid-speak these days would call 'the fiscal modalities.' For although just about anybody could peg claims just about anywhere (provided they first discovered something), the Chartered Company demanded a 'free ride' of fifty percent of the profits.

Worse, individuals and syndicates were not allowed to mine. Only public companies were—in which the Chartered Company would get their free 50% in the form of shares. It was another indication that some vast, rich discoveries were confidently expected; the anticipated scale of the mines would surpass those of the Witwatersrand and small investors were not welcome.

All poor together

There was yet another feature that suggested this. It was drawn from American laws, but fiercely argued against by Hays Hammond, his chief American adviser.[51] This allowed claim-holders to assert 'extra-lateral rights' if a reef dipped outside the vertical limits of a claim. It was an incentive to rapid mine development, because it reduced the need to establish the geological structures at depth, regardless of any adjoining claims. However, geology is usually not so simple; reefs can split and join, be displaced by faults and run in parallel. So there is often plenty of room for argument as to whether a reef structure being followed down is indeed one and the same throughout. Hays Hammond knew that the rule had led to many law-suits in the United States, and indeed in Southern Rhodesia it was to result in a marathon case in the 1920s, involving the country's richest mine, the Globe and Phoenix. This was eventually only resolved by a decision in the House of Lords.

But from Rhodes' viewpoint it was all one. Why worry about the next-door operation cutting in to somebody's reef when his Chartered Company held effective control of every mine in the country anyway? The prospectors were enthusiastic about the rule, that was the main thing. As long as they kept looking, it could not be long before King Midas himself would be outshone.

To cut a long and unhappy story short, if the gold in the Rand had indeed come from Rhodesia, it was long gone from that country. Indeed, on sober reflection—understandably a difficult activity in those urgent early days—it seems most unlikely that enormous amounts of gold would have been washed over such a great distance. Twenty, even fifty miles perhaps. But six hundred? The Lydenberg goldfields, or those of Tati, or Eersteling, might just be the fragmentary remains of the mother lode, but the likeliest remnant, if there are any at all, is Fred Struben's Confidence Reef, on the greenstone belt known as Muldersdrif, abutting the conglomerates. Nothing was found in Rhodesia remotely as good as the Rand.

As a result, although something like 65,000 claims had been pegged by 1895, few discoveries big enough to justify share flotations had been found. More, the regulations had not merely severely inhibited mine development, they caused the wrong sort of mine to be built. Companies raised money on the stock markets of Europe and America for big mines; great sums were spent on roomy shafts and massive mills. Yet the mineral resources on which these developments took place were usually too small to justify the capital expended, and this (along with much Buckreef-type mismanagement) led to the ruin of many of these ventures.[52]

Rhodes' gamble for wealth from 'his North' had failed, and from his response flowed much twentieth-century history. For now there was only one other throw he could make, one other place where patriotism and business could be usefully blurred into exciting synergy. The limitless riches of the Rand.

After all, heaven knew that there was provocation enough against the *uitlanders*, many of them British, from the Afrikaners. Despite the economic power their gold mining skills gave them, and despite their providing the government with most of its revenue, the *uitlanders* could not vote except after many years of residence. Monopolies and concessions for mining essentials (the high cost of dynamite was a particular affliction) were granted only to Afrikaners or their kinsfolk, such as the Dutch. Much of the *uitlanders'* taxes were going into what seemed to be excessive military expenditure. It all sounds rather like latter-day Zimbabwe; perhaps fortunately nowadays the British Government is profoundly indifferent to complaints of discrimination and corruption from whites in its ex-colonies in Africa.

The Jameson Raid, should you need reminding, was an attempt to overthrow the Republic of the Transvaal in 1896 and replace it with a government favourable to the *uitlanders*. It failed miserably because the intended beneficiaries got cold feet. This adventure, which, assuming that Rhodes was as schizophrenic as he appeared to have been, may be envisaged as a *folie a deux*, utterly destroyed his South African political career (he was Prime Minister of the Cape at the time). It also was the start of the destruction of much else:

> In the year 1895 I had the privilege, as a young officer, of being invited to lunch with Sir William Harcourt. In the

All poor together

course of a conversation in which I took, I fear, none too modest a share, I asked the question, 'What will happen then?' 'My dear Winston,' replied the old Victorian statesman, 'the experiences of a long life have convinced me that nothing ever happens.' Since that moment, it seems to me, nothing has ever ceased happening ... I date the beginning of these violent times in our country from the Jameson Raid of 1896.

So wrote Winston Churchill in his review of the events leading up to the First World War, taken from his work *The World Crisis, 1911–1918*. Improbable as it might appear now, up to that point war with France seemed more likely than fighting the Germans; there were numerous flash points with the French, particularly in Africa, where parts of the Sudan, Uganda and Nigeria were all in contention. But the Jameson Raid led to a telegram from the Kaiser expressing Germany's support for President Kruger and the Boers, and from that time on there was a growing polarization between the British and German empires.

Rhodesia was fortunate to survive the vicissitudes that followed the Raid. The enterprise had drawn heavily upon the few European policemen there, leaving just forty behind to control the entire country. Partly as a result of this, severe rebellions by the Matabele and Shona tribes (the 'First Chimurenga' or Liberation War) came near to closing it down; 'Imperial' troops had to be drawn upon in the end. There was also an outbreak of rinderpest that eliminated almost all draft power at about the same time. Finally came the disruptions of the Boer War (1899–1902), whose origins can, again, be traced directly to the Jameson Raid.

Rhodes was dead by the end of the conflict; his overstrained heart had finally given way. It may have been the right time to go; the coming century was to be unfriendly to Empires and their builders, and the antagonisms and conflicts that had flowed from his reckless scheme may have come to be identified with him. As it was, his main legacy was a system of generous scholarships (initially restricted to candidates of Anglo-Saxon stock) that have given many of the great and the good that benefited a lingering

Ruined by drilling

affection for the old rogue. Mark Twain summed this attitude up—'I admire him, I frankly confess it, and when his time comes I shall buy a piece of the rope for a keepsake.'*

• • • •

When Rhodes died in 1901 the annual output of the gold mines in Rhodesia was only 194,000 ounces, or about six tonnes. By comparison, in the nine months before the outbreak of the Boer War in October 1899, the Rand had produced 3,600,000 ounces, or about 125 tonnes. Yet in the final analysis he had done the future of his country an enormous favour with his mining law. This was because, stripped of the crippling 50% free carried government share that it originally demanded, the legislation was a perfect match for its diverse but complex and lean geology.

In 1903 the Chartered Company's directors, rid now of the need to do the wishes of the colossus that had created it, allowed mines to be built and operated by any form of organization, not just public companies. Their 'free ride' to the revenue from mining was reduced to 30%, and then to nothing a few years afterwards. Thereafter an aspirant miner, people like my father fifty years later, whose capital limited them to driving an adit on a reef going into the side of a hill and setting up a second-hand three-stamp battery to recover the gold, was able to get into the business of mining.

By the start of the First World War nearly half of the gold produced was coming from several hundred smallworkings. Smallworker activity peaked during the Great Depression when financial distress led to many new entrants, such that in 1934 there were 1,600 small gold producers. It may be noted that at the same time there was a similar boom on the other side of the world, back on the mother lode in California, when thousands of unemployed men manned an estimated 3,000 smallworkings and produced more gold than had been seen since the peak output of the forty-niners.

* *More Tramps Abroad*, 1898.

All poor together

Rhodes' desire to get things moving as quickly as possible also resulted in two other useful features of the present Zimbabwe mining law. Both were designed to prevent a leisurely or fatuous approach to the mineral resources he was after. First he made the discovery of a deposit as a pre-requisite for title and second the obligation to maintain its validity by work or payment. It was not and is not possible arbitrarily just to peg a claim and sit on it for ever more.

Whatever the details, the point is that the system worked, and still does. Little Zimbabwe, with its five hundred or so little mines producing thirty-five or so different minerals, has more than all the rest of Africa (outside South Africa) put together.

Chapter Ten

GOLD AND GOD

> *IN THE NAME OF ALLAH THE BENEFICENT, THE*
> *MERCIFUL ...*
> *Oh ye who believe! Devour not your wealth among yourselves*
> *by unlawful means, but earn by trade based on mutual*
> *consent. And kill not yourselves. Surely Allah is merciful*
> *towards you. Truthful is Allah the Magnificent!*
> —Quote from the Koran heading Dar Tadine Tanzania's
> Morabaha gold certificates, 1985.

The segmentation of the mining business into promoters, developers and miners has evolved that way with good reason. The prospective buyers of a mineral title always carry out, or should always carry out, a due diligence. In a succession of such transfers there is therefore a succession of independent checks on the probabilities (not certainties, remember; this is a casino) of the claims of the sellers being valid. During these investigations the pretensions of exploration companies like Gargantuan Resources Inc. get suitably deflated. Attempts to short-circuit this, like Cecil Rhodes' combining all three in his Chartered Company, usually come to grief. The type of cool, indeed cynical, evaluation needed at each stage by the mine-hunter when galloping from a scent to a view to a death can easily be warped by corporate and individual pride amongst his advisers. Here is an example; I was one of the advisers.

Dar Tadine Tanzania was an Arab, nay, Islamic, company, that set out to find, develop and mine gold deposits in Tanzania in 1984. It had a genius of a promoter, Dr Ibrahim Moussa Kamel, an Egyptian, who raised twenty-five million dollars as a first step. He was tall, immaculately suited, with a great presence and a penchant for private jets. His sole false note, and then only for Anglo-Saxon sensibilities, was a tendency to wear his overcoat slung around his shoulders in the manner of a certain type of ageing central European émigré.

The exhortation to Allah above, devout yet opaque, serves to

All poor together

illustrate his style. He was a high flyer in the business world of the Gulf and had been a vice president of a bank there. There had evidently been some disagreement about his dealings in precious metals—nothing culpable, you understand—and he left the bank armed with useful contacts amongst middle-level entrepreneurs and contractors in the region and, I can only guess, a knowledge of the recondite world of gold futures.

These Gulf businessmen were a good target market for an entrepreneur, but what project would fit their ambitions? It may be hard to imagine this now, but an important one at the time was to get out of U.S. dollar holdings and into something more stable. Since 1971 businessmen in the Gulf had seen the gold price soar as the dollar weakened. It is important to remember throughout this story that it was received wisdom amongst many wealthy persons at that time that gold would continue to strengthen against the U.S. dollar.

The other objective of this clientele—aspiration is a more seemly word—was to be, and to be seen to be, good practising Muslims. Orthodoxy had commenced its fierce ascent only four years before with the Iranian revolution, and it was not a desirable thing for a merchant or contractor in the Gulf to be thought irreligious.

Gold and God; a powerful combination. How to fit them together?

It so happened that there was a marvellously equipped airborne geophysics exploration company that had got into a spot of trouble in Tanzania. This was Geosurvey International, whose little yellow helicopter had arrived at Buckreef with the hapless Mr Vergympf, and its problem was, very simply put, naïveté. For five years, from 1976 to 1981, it had sweated over a contract for the Tanzanian Government. Its aircraft flew over a million miles, backwards and forwards just four hundred feet above the ground, to produce what must be the definitive record of the magnetic forces lurking under the surface of that country. But at the same time during that period Tanzania was getting poorer and poorer. There were droughts, the war against Idi Amin, the collectivization of the peasants, corruption, mismanagement and disillusioned donors. Anyway, when the time came, the country could not, would not, pay the DM.70 million—U.S. $25 million at the time—that it owed Geosurvey International.

My sympathies are entirely with that company. I did some work for the Tanzanian government twelve years after that— indirectly a World Bank job, as it happened, but the money was already in the Tanzanian central bank and the project was being run by Tanzanians. Yet even though the Tanzanian version of financial *glasnost* was in full swing, I didn't get paid. No, I tell a lie. I did get paid in the end, but from Washington, not Dar-es-Salaam, eight months later, when it became clear that the Tanzanians were never going to do anything about my invoices. No worry about the job I had done, that was fine. Just—no money.

So it was with Geosurvey. Technically they did an excellent piece of work, which is being used, and will be used, for decades to come. But as to payment, all they received was a wide range of rebuffs. Based on my own experience I would guess that they included evasion, airy unconcern, professed ignorance and finally, all excuses exhausted, dull obstinacy.

At that time, I was told by persons who had seen it with their own eyes, there was a meeting every morning at the Ministry of Finance to decide who got paid that day against the dribs and drabs that were coming in. No doubt the usual hierarchy got priority—the oil companies, the arms suppliers, the spare parts for Air Tanzania, the strip for the national football team, the four-by-fours for ministers and so on. Commercial creditors came last, and it was best, if you absolutely had to be a creditor of Tanzania, to deal in small amounts. Twenty-five million dollars? Why (the minister might well have said), we could build another two Buckreef gold mines for that (it being about that time when great things were still expected of that calamity). Geosurvey International could wait.

Fate had it that the owner of Geosurvey International, a delightful German called Peter Gollmer, was to meet the good Dr Kamel, perhaps at the latter's offices in Geneva. And perhaps Peter Gollmer, a man with a twenty-five million dollar hole in his bank balance, had given out, as policemen say, that his aeroplanes could find gold mines.

All poor together

This is perhaps the place to reiterate that, certainly at the time this happened, no gold mines existed that were found by aeroplanes. There were, however, two spectacular finds by ground geophysics, quite a long time ago. First, in 1930, a magnetometer was used to detect the magnetic shales above a western extension of the Witwatersrand conglomerates, giving rise to another crop of Rand mines. Then, in 1936, the Free State gold-fields were first detected using gravitational changes (there were no magnetic shales in that area of the Witwatersrand basin). The latter, when they came into production after the Second World War, were the main reason that South Africa's output went up from 360 tonnes in 1950 to just over 1,000 tonnes in 1970.

But both these earth-bound geophysical discoveries were on the Witwatersrand formation, a geological freak in itself. Some hydrothermal gold mines—a very few—have also been found by ground geophysics, more by geochemistry. And certainly, yes, after a mine has been found, geophysics, even the airborne variety, is valuable in building up a picture of what is hidden down there. Remember from a couple of chapters back that I mentioned that Geosurvey International had found some very good geological indications in its five years of airborne surveying in Tanzania. The next steps were first to use a helicopter—the self-same yellow helicopter—to increase the geophysical detail of the anomalies found, and then to test hundreds of soil samples above them for gold. However, when the company closed in for the kill—selecting eight places where the geological, geophysical and geochemical information all combined positively to shout for attention—the drill holes only intersected massive ore bodies of almost worthless iron pyrites, with negligible gold values.

No matter. Peter Gollmer was looking for money, Ibrahim Kamel for gold mines. It seemed a marriage made in heaven, or perhaps paradise in this case. Peter's aeroplanes would fly about with long aerials on their noses or boxes of gadgetry dangling down underneath, finding gold mines, and Ibrahim would produce gold for his investors from those mines.

Ibrahim went to work on the Tanzanian Ministry of Minerals, or rather on the minister, the permanent secretary, the commissioner of mines and the general manager of Stamico (Pat Carter, the party animal). He emerged with concessions totalling seven thousand square kilometres, covering the best areas of the greenstone belts of northern Tanzania.

At that time the law required that the government have a majority holding in any mining company, so in a neat move (there was to be a bewildering array of such neat moves, perhaps that was the problem) the concessions were held to belong to Pat Carter's Stamico and DTT were contracted to that decaying organization to produce gold on the concessions by a production-sharing arrangement. In a tit-for-tat DTT was to be exempt from all duties and taxes.

So the investors in the Gulf would get their gold for their U.S. dollars. But how could they get God as well? Gold mining and Islamic principals? Could there be a connection?

Yes, but be prepared to learn some Arabic, some *Sharia* law and to think laterally. First, the word for a venture is *Modaraba*. The organizers are the *Modareb*. The subscribers are the *Arbab El Maal*. *Dar Tadine Tanzania* (DTT) was the executing unit, the one that found and built the mines—and while the prospectus described it as wholly-owned by the Modareb and the Arbab Almaal, it was carefully specified that DTT was wholly-controlled by the Modareb.

The subscribers were invited to put up, initially, 25 million dollars for exploration, perhaps coincidentally the sum that was owed to Geosurvey by the Tanzanian Government. This was to be put into an Islamic, and thus non-interest bearing, account held by the Banca Della Svizzera Italiana in the Bahamas. Fortunate bankers. The 25 million was to be converted into shares of $25 apiece, non-voting ones; voting was to be limited to the Modareb. Unfortunate subscribers. However, they could console themselves with the knowledge that 'the Modareb undertakes in fulfilling all its obligations under the Modaraba Contract to respect the provisions of the Glorious Sharia under the provisions of a Religious Supervisory Committee composed of two leading scholars of the Islamic nation and a third member representing the Modareb.' From the Muslim viewpoint the important thing was

All poor together

that it had been pronounced, perhaps by the 'two leading scholars', that this investment qualified as a *Zakat* or tithe, because of the way in which the scheme was to benefit Tanzania, a poor country with ancient and cordial Islamic connections. On the coast, that is; inland around Lake Victoria the ancient connections were somewhat less cordial, as I discovered.

To qualify for *Zakat* status, Tanzania was to get a third of the gold produced—the production-sharing contract with Stamico. In addition, of the initial U.S.$25 million, U.S.$10 million was to be available as a loan to Tanzania, to be repaid at a rate of 25% of the gold it received—but not in metal; it was to be paid in its U.S.-dollar value and this money distributed to the subscribers, the *Arbab El Maal*. The gold price, remember, was going to go up. Everybody said so.

The subscribers had, through their payments for what was known as the Exploration Offering, qualified for the next stage, which was the real earner, the Extraction Offering. Each mine found was to be funded separately by one of these, although as far as I know the Buhemba tailings plant, whose story follows shortly, was the sole offering achieved.

The prospectus showed how the system was to work:

> DTT is able to offer for present payment Islamic Dinars to be delivered in the future pursuant to an Islamic Morabaha contract at a gold price lower than the international price of gold today. Financing the costs of production through this Islamic Morahaba sale eliminates the cost of forbidden interest ('Riba') from a mining company's cost of gold production.
>
> For each U.S.$862.00 invested in a Morahaba certificate, the subscriber, Allah willing, will receive twenty (20) Islamic dinars, each with a gold content of five (5) grammes (0.1608 ounces).

So, payment now would lead to the lender receiving more gold in the future than that payment would realize today. The timing was brilliant. When the Morahaba were launched in September 1986, the gold price had climbed from $326 in January when the prospectus was put together, to $442 an ounce. Or, to look at it in the same way as the rattled rich in the Gulf, the U.S. dollar had declined by as much in gold terms.

The plan was that certificate holders would get their Islamic dinars every quarter over the next five years. So by putting up $862 for 100 grams of dinars now (a price equivalent to U.S.$ 268 an ounce) you would get back gold worth, oh, almost certainly over $400 an ounce, in twenty tiny bits (a five-gram coin would be about twelve millimetres diameter and nearly three millimetres thick) every three months to 1991.

Of course, you had the option of just leaving these to accumulate, or of selling the bearer coupons that signified your ownership to somebody else. That would be, it was hoped, what most people would do; a market in such coupons would obviate the need to knock out all the dinars spoken for, so giving DTT a useful cash float.

This turning of cash into tradeable promissory notes for future gold was the clever key to a scheme with a host of clever, if not complex, angles, that hallmark of an Ibrahim Kamel deal. What the doctor planned was not a deception; banks do it with your money all the time, handing it out to borrowers on the assumption that you and the rest of the depositors won't all come in simultaneously to draw it out. The important difference was that this would be cost-free money; there was no interest to be paid by DTT; Sharia law forbade it. You were guaranteed that in the event of *force majeure* you would get your investment back, no more, which was presumably secured by a gold price hedging arrangement.

That was the deal, Allah willing. The note of crossed-fingers-behind-the-back was pertinent; the gold had to be found first.

· · · ·

The illusion that gold mines still lay awaiting discovery somewhere in the mass of airborne geophysical data accumulated by Geosurvey was swiftly shattered. Certainly, by the time I became involved, in late 1984, conventional geologizing was the thing, and I found myself touring the old tailings dumps and dams with a couple of Geosurvey geologists, looking for gold the old-timers had left behind that might make quick pickings for DTT.

The answer, we found, was 'not much'. Tanzania's fifty or so gold mines (all of which had closed at or after independence) had not been rich, and only three or four had been of a reasonable size. Most of them seemed to be the type of smallworking my father had lost money in. The best opportunity to get gold quickly from tailings appeared to be at a big old mine called Buhemba, in a broad valley near the eastern shore of Lake Victoria, close to the Kenya border and about eighty kilometres from the nearest town, Musoma. Buhemba was said to have something like a million tonnes of tailings in a huge dump.

A dump, not a dam. Nowadays gold ore is invariably milled up fine and cyanided in big agitators in one fell swoop. The sloppy pulp that comes out at the end of this process cannot just be piled up, but has to be fed into a dam where the particles can settle out and the water be recycled.

This 'all sliming' process consumes much energy and was introduced on the Rand in the 1930s when big new power stations brought the cost of electricity to reasonable levels. Prior to that the ore was ground to a relatively coarse size, and the resulting particles put into 'sands tanks' where cyanide solution was pumped in to percolate through. The tailings from this process were simply damp sand, and this was merely shovelled out and dumped. It was a route that was still favoured in remote mines after the Second World War, such as Buhemba.

But Buhemba was now an army camp. This was unsurprising; abandoned mining villages provide excellent out-of-the-way accommodation for Africa's big armies and we had already been to one such mine, Kiabakari, where a parachute battalion was based. We had a letter from the Ministry of Defence in Dar-es-Salaam, which had worked well enough there, and we were let in to look over the tailings dam. A single panning was enough to confirm what we already from the literature of colonial times;

there was no useful gold in the Kiabakari tailings. But when we arrived in Buhemba we were refused admission; this was a national service camp which worked (for sound political reasons, if you put yourself in the *Mwalimu*'s shoes) under a completely different ministry.

There were, at that stage, three of us: Tony Roberton, Howard Bills and myself. The man in charge, a worried captain, was pleasant enough but insisted he could not make a decision until the colonel came back from Dar-es-Salaam, four or five days later (Allah willing, we thought). It was fairly late, so we set off down the valley to camp a couple of kilometres away, en route for the Mara Mine, our next stop.

Not far enough; we were rounded up by a sergeant and a small group of conscripts and told to come back to Buhemba (you can never underestimate paranoia). Here the captain politely told us that we would have to wait there until he had radioed Dar-es-Salaam or, alternatively, the colonel arrived.

Our new accommodation was part of what had once been the mine club (the steel base in an erstwhile billiard room gave it away), and we spread our sleeping bags on the floor of what might have once been the library and resumed preparing supper. Because they anticipated difficulty in getting food in Tanzania, Tony and Howard had brought out a selection of TV dinners from England, and it was on these unlikely meals that we survived during our house arrest.

Sharing the building with us were two melancholy individuals from, of all places, Guyana in South America. One of them had the job of showing Tanzanian national service conscripts how to make the drums for a steel band. The other was there to train them in their use. They had been in the camp for about eight months, and were now two months overdue to return home, but they seemed to have been forgotten. No word had come from Dar-es-Salaam about flights and so on, and they were getting desperate. Perhaps the money needed for their return had not surfaced in those morning meetings at the Ministry of Finance.

They were busy enough; for lack of anything else to do they were training a fresh batch of recruits. The drum factory was next

All poor together

to the house, the rehearsal shed opposite it. The cacophony started at seven sharp in the morning and lasted into the evening.

After a couple of days of this we lobbied the captain with some desperation, and were allowed to quit our house during daylight to inspect—not sample—the tailings. We were watched, both officially and unofficially, by the olive-clad conscripts who were milling about the place, but I managed to pan the tails, if not sufficiently surreptitiously, at least without being halted. They didn't show up much, the colours were sparse and very fine, and there was a fair amount of pyrites in them as well.

That was a bit of a worry. Over the years in a tailings dump the pyrites slowly oxidize away, freeing up locked gold for recovery in cyanide solution. Not just once, either. The main tailings dam at the Cam and Motor mine, once Zimbabwe's largest, has now been retreated three times over the last sixty years, each time getting out a smaller fraction of the remaining gold. On the last occasion it was about a third of a gram a tonne.

We made one valuable discovery. There was a flat space of about two hectares overgrown with grass so tall that all we could see was the blackened, rusting rims of steel tanks projecting above it. This had been the plant area, and after fighting our way through the vegetation to them it turned out that they were the sands tanks, where the gold was extracted by cyanide percolation. Now if these could be patched up, then we would have a cheap way of getting the residual gold from the tailings, whose white hills loomed all about. Cyanide could be pumped up from a sump, allowed to drain through the tailings in the tanks and the pregnant solution treated with carbon or zinc to precipitate the gold. The barren solution would return to the sump; one smallish diesel pump would do for the whole system. We wouldn't even need electricity.

The captain came around regularly for the first day or so to say that he had been unable to make contact with Dar-es-Salaam. After that he avoided us, and we were left to clamber about the dumps as we pleased. From the top we glanced wistfully at the gap in the hills at the end of the valley that would be our escape route, but short of ramming the boom at the entrance to the camp, we were evidently doomed to stay until the colonel arrived.

Eventually he came, red-capped, red-tabbed, small and portly, driven in a clean, newish dark green Land Rover. We were summoned to his office. He knew all about us; indeed he told me that there was to be an aeroplane at Musoma later that day to collect me. Well, yes, that was right, I was supposed to be at Musoma now.

'Can I take some samples?' I asked. 'They are expecting me to do tests.'

How much? he wanted to know. A few kilograms, I told him. The gold in it (I calculated furiously) would not be worth more than a couple of shillings (about ten pence at the time).

He digested this figure. There was a silence. Then he rapped out a series of commands and we were escorted briskly out and taken up to the tailings. A couple of conscripts were detailed off to come with us, as did the captain, by now anxious to be helpful. We walked over the top, grab-sampling here and there while above us grey-black clouds were bellying and rumbling; the rains were starting.

We filled half a dozen plastic bags—enough for a start, I judged—and returned to say goodbye to the colonel. As we drove out of the gate the heavens opened. Standing in a doorway, watching us go with yearning eyes, were the Guyanese; I hope they got home in the end. I didn't seen them again.

The twin Beechcraft was already on the apron when we arrived at the little town of Musoma, on Lake Victoria, the rain spraying off it. Inside Malcolm Hooper, the project director, was waiting.

We splashed across and crouched in conference in the tiny freight compartment in the back while he questioned us.

'Ibrahim needs to get some production as soon as possible.' Malcolm was on first-name terms with the great doctor. 'What about Buhemba?'

'We have some samples; I can tell you in a week's time if it is any good.'

'But what do you think now?'

I told him it was the best by far of all the dumps we had seen up to then. Because the old plant was still there in the long grass, we could probably get started pretty cheaply. Quickly, too.

All poor together

'How quickly do you think? Ibrahim hopes to get started next month.'

Next month, eh? And this is Africa. Ibrahim was asking for a lot.

It was arranged that I would hustle through the tests. If they looked good we would press ahead with rehabilitating the plant. At the same time the whole dump would need to be sampled properly, not just the scratchings we had taken from the surface.

Malcolm got out with Tony and Howard and they raced back to the streaming Land Rover; he was to go to Buhemba to see for himself. I stayed on board until the rain slackened, when we took off for Nairobi. We were half-way there when I realized that I had left Tanzania without any customs or immigration formalities.

• • • •

Four months later I was at Mwanza airport, about two hundred and fifty kilometres from Buhemba. Here, under an electric blue sky, a clutter of light aircraft and a big private jet stood shimmering. Inside the latter Dr Kamel was entertaining a number of potential investors. These gentlemen were Saudi and Kuwaiti businessmen, the *Arbab El Maal*, complete with gleaming white robes and the *qalifeh* head-dress, come to see for themselves the first gold to be produced at Buhemba.

The jet resembled a short, tubby version of a 737. It was called a Challenger and Dr Kamel believed it to be the largest private jet in Africa. In fact Sol Kerzner, the South African magnate of Sun City fame, had one as well, but it probably would not have been very profitable to point this out.

The jet belonged to Geosurvey International, and Peter Gollmer had used it to flit about the globe in. Maybe it was the Challenger that had decided Ibrahim Kamel to take over the company. Certainly he was flitting about the globe. One of the pilots told me that they had on occasion taken off with a great show of urgency and elan, climbed to 35,000 feet or so and stooged about until the Doctor had decided where to go. I was to discover that this was something of an occupational hazard for the pilots of private jets;

the very availability of this gleaming magic carpet brings with it a need to sweep off urgently somewhere—anywhere—preferably not unnoticed by eminent persons on the ground.

Mwanza, with its tarmac airstrip maintained for the dwindling squadron of Russian Mig's that crouched in bunkers there, was the only airport in the region where the Challenger could land. The gaggle of light aircraft had been brought in from Geosurvey in Nairobi to take us to the airstrip at Buckreef, the nearest to the other concession areas allocated to DTT.

Here I greeted old friends before we set off in Land Rovers into the bush, to see the activities of the artisanal miners in the DTT concessions. (Dr Kamel had a plan for them as well, one in which I was to be involved.) Dr Kamel, clad in a safari suit, strode ahead. The robed Arabs tripped cautiously through the bush behind him, peering impassively into the shafts and pits of the bemused locals. Unkindly, the sun silhouetted the outlines of the visitors under their impressive raiment, revealing them to have beer-bops as substantial as any Boer farmer.

From Buckreef we returned to Mwanza, where the investors went into the Challenger for refreshments. Its auxiliary turbine was roaring as it struggled to keep the air-conditioning inside suitably icy. There was a story that when the Challenger first landed there from, shall we say, Kuwait City, the door opened, the ladder extended itself and a squinting aide came cautiously down in the blinding sunlight. In his hand he carried all the passports and a large sum in U.S. dollars. He peered around at the supine baggage handlers and scratching, scruffy officials and picked the one there who was in full, impressive uniform. Into this individual's hands he pressed the money and the documents and retreated back inside to report, leaving behind a perplexed airport fireman.

On this occasion we, the dozen supernumeraries, pilots, geologists and so on, squatted in the shade of the wings of the light aircraft. At intervals the odd Egyptian gofer and flunky would walk down from the Challenger, drinking from moisture-beaded cans and bottles, grinning smugly. We looked at them with disdain threaded with great envy.

There was a Swiss free-lance photographer with us, a girl who

All poor together

had been in the Angolan civil war, amongst others. She marched up to the steps and vanished inside, reappearing with an armful of bottles of the Tanzanian Double Cola drink. They were deliciously cold, and vanished rapidly.

For me there seemed to be a nice dilemma in etiquette at this point. Tanzania was one of numerous African countries where the bottles were more valuable than the contents in a very practical sense; you could not get any more Double Colas unless you brought back the equivalent number of empty bottles. The same rule applied, with even more force, to Safari beer and pilsener; there was no hard currency available for more bottles.

So, to the surprise of my European companions, I collected the empties together and set off for the steps of the Challenger. A blast of icy air met me as I started to climb. With a sense of fitness I found an immaculate European air hostess in the doorway. I indicated the bottles.

'Do you want these back?'

'Oh heavens, no!' she exclaimed, disdainfully.

So there is quite a useful bit of information. If you own a private jet, you don't worry about the empties.

* * * *

Eventually the white-clad figures streamed out of the jet and were helped by the pilots into the waiting light aircraft. We flew in a roaring stream to Musoma, where more Land Rovers were waiting to take the party to the resurrected Buhemba plant.

This was my moment. We were to smelt the first gold. The place had been transformed. The vegetation that had covered the plant area had long gone. The tanks had been repaired and made watertight, a ramp had been built to allow tractors to tip into them and the rails-and-cocopan system beneath them had been recommissioned so that the cyanided material could be sent back on to another, new, dump.

The piping and zinc boxes had been put in place and there was now a Filipino metallurgist there, Jerry Alo, to run it. He and I had worked together to commission the primitive system. He had washed off the black slime that contained the gold from the zinc shavings, and the residual zinc in this had been dissolved away in sulphuric acid. The even blacker residue that was left had been roasted to paleness on a tray over a fire, and suitable fluxes mixed in. Judging from the solution flows and the tests—we had no assay lab—we thought we had four or five ounces. Maybe more.

Malcolm had already asked me if I would like to have a bit of artisanal gold handy to ensure that we made a decent showing. I put on a po face and said good heavens, of course not. That was very foolish.

For already things had started to go wrong. There were no firebricks to use in the simple furnace we had built in an old 200-litre drum and the refractory cement needed to seal them in hadn't arrived either, so we used ordinary local bricks backed up with sand. The blower that was to give the forced draft had not arrived—or rather one had arrived that was about fifty times too big and would present huge difficulties in transporting and setting up. We improvised by connecting up to the outlet from a compressor.

We lit up the local charcoal that was our fuel and started up the compressor. The group of investors gathered around, a rain of sand came up from behind the bricks over them and they started back in indignation. I bellowed for the compressor to be stopped and we made some hasty modifications.

On restarting rather less sand leapt into the air, and we decided to continue. However, the wretched furnace was not really hot enough, and after forty-five anxious minutes we poured a rather sticky melt into its conical mould in the middle of a tense circle of visitors.

After a minute or so I inverted it. The guests pressed closer as out dropped a cone of green slag with a small, brownish tip. I blasphemed silently. Where was the gold?

Nothing to do but press on. I tapped the tip sharply and it fell off with a satisfying thud. So it was metal, thank heavens. I dropped it sizzling into the bucket of water standing ready and after another minute felt down and pulled out the still warm cone, chocolate brown and lumpy, and displayed it to the throng.

All poor together

'It's gold,' I said. There was a ripple of barely polite applause. Somebody said, 'It looks like copper.' The dreadful thing was passed from hand to hand and eventually to Ibrahim Kamel, who was very nice about it — 'Our first output; there will be much more after this' — and handed the thing back to me.

I had a job to do with it. I said goodbye to Jerry Alo and the rest and set off for Musoma again.

• • • •

Geosurvey International had its real base at Wilson Airport at Nairobi, Kenya's light aircraft centre. Here there was a huge hangar in which all thirteen of its aircraft could fit. They were an impressive sight, sparkling clean, standing on an immaculate white floor.

Workshops and offices were built into the hangar. Here, in the big board room, I had first met Malcolm, who had arrived from the Philippines straight from a Manila hotel fire. He had lost all his clothes apart from those he was wearing and not long prior to that he had survived a light plane crash in Saudi Arabia. He had been in charge of a mining project in Saudi, and had been recruiting for it in the Philippines. Dr Kamel had swept him up from Manila in the Challenger in the hope of getting him to take over the project.

I don't know what his arrangements were with the doctor, but they would have been complex. I was offered an improbable deal involving payment of my fees through an irrevocable letter of credit, which I wonderingly turned down for more conventional payments.

Amongst the services provided by Geosurvey was a laboratory. David Sims was the chief chemist at that time. It was about six in the evening; he was anticipating working late, which was lucky. I produced the thing I had in my pocket, which had not, of course, been declared to customs in either Tanzania or Kenya. His eyes widened.

'Is that it?'

'That's it.'

The lump was actually a matte; there was quite a lot of sulphur

Gold and God

in it. The brown colour came from copper and God-knew-what-else. We had to produce pure gold from this unhappy mixture.

It was necessary to dissolve out the impurities, and to do this we had to break it up into fine particles. There was not enough sulphur in it to make it friable enough to crush, so we melted it down again and granulated it—running the molten metal directly into a stream of water so that it formed thousands of tiny spheres.

This was a rather perilous operation, as hydrogen would be formed from the impurities during granulation and a false move might not just have blown the gold all over the lab but might have done the same for us. I had been working on a nickel smelter once when somebody was killed by a granulation explosion.

To our enormous relief the operation was noisy but uneventful. We carefully collected the thousands of little brown granules from the bottom of the sink we had used and slowly emptied them into beakers of nitric acid sitting on hot plates in fume cupboards. Poisonous brown fumes of nitrogen dioxide billowed out as the copper and so on was dissolved, until we were left with containers of gently simmering blue liquid at the base of which was a black residue. That was the gold.

We filtered and washed it and then put it in a crucible and melted it in an furnace. To my indescribable relief, when we took it out pure molten gold, a rich yellow, was wobbling around inside.

But that was, in retrospect, the easy part. A steel mould had been made in Switzerland to enable us to cast gold Modharaba coins from the first production. Not measly little five-gram dinars but Kruger Rand equivalents, one-ounce coins of over thirty grams. The doctor was due to meet with President Alix Mwinyi of Tanzania (it was shortly after the *Mwalimu* had chosen to enjoy power without responsibility by standing down as president but remaining as head of the party) the very next day to hand one over as a keepsake.

We could not do it. Gold has a melting point of over 1,000 degrees Centigrade, and we were able to achieve that. But even though we heated the metal mould to warping temperatures, as soon as we poured the gold into it, the metal froze, giving us pathetic bits and pieces of the real thing.

All poor together

With the perfect wisdom of hindsight we should have had a ceramic mould, or a press to stamp the coins out. In the event I had the humiliation of going out to Jomo Kenyatta Airport, waiting there for the great man's jet to arrive and instead of producing several one-ounce coins, handing over two small smooth lumps of very, very pure gold.

He nodded at my explanation, slipped them into his blazer pocket and strode off with his little train of flunkeys trotting after him. Not long after I saw a photograph of him on Zanzibar, in the act of handing one of these embarrassing blobs over to President Mwinyi.

• • • •

Even if the *Arbab El Maal* had been right and the price of gold had continued to climb, it was difficult to see how the DTT venture could have made much money. Tanzania may have lifted all duties and taxes for the venture (and as always happens, the customs and tax departments were only very belatedly informed of this), but the effective royalty of 33% from the production-sharing contract would have been a heavy load for a very rich, big mine. Tanzania, like Zimbabwe and unlike West Africa, has never been a country of big, high-grade gold mines.

In the event, the massive exploration effort that was made—at one time there were over sixty expatriates on the payroll—was directed at the wrong target. The numerous artisanal reef discoveries were ignored in favour of a search for eluvial and alluvial gold. This was a strange objective in a country with a limited depth of weathering and a flat topography. Predictably no significant deposits of either type were found.

As might also be foretold, the U.S.$10 million lent to the Tanzanian Government was not repaid. Nor were rich mines discovered. In 1988 the government reduced the concession areas by 50% and the base camp in Mwanza was briefly expropriated over a legal dispute on land ownership. A scheme to buy gold from artisanal miners (those Islamic dinars, Allah willing, were falling due) led to a parliamentary inquiry into allegations of gold smuggling and illegal trading. And so on.

The parliamentary inquiry, incidentally, failed to turn up any substantive evidence for these accusations, although much was made of indications of DTTs 'trading with the enemy' (these were Kenya and South Africa at the time). However, it served to strengthen my own suspicions of a broad distaste in upcountry Tanzania for the overt Islamic nature of the venture.

The (at that time illegal) artisanal gold mines in the DTT areas were, of course, already well served by a network of gold buyers, few of whom seemed to be Muslim. As elsewhere in East Africa, Asians of the Hindu persuasion were the principals in the trade. Dr Kamel had the bright idea of using some of Egypt's great army of underemployed graduates to form his gold-buying teams, and while their behaviour was, as far as I know, impeccable, they were up against powerful competitors who would have no hesitation in playing the religion card.

There were other sinister echoes to be harped upon. Arabs were associated with the slave trade, which had weighed heavily upon the peoples of the Great Lakes region until only a hundred years before. Then, ninety-five years ago, Emin Pasha's raggle-taggle army of Egyptians had accompanied H. M. Stanley, their 'rescuer', through the area. They were on their way to the coast for repatriation, a hungry and undisciplined horde. Comments by local people to me at the time of DTTs gold-buying suggested that these memories still lingered. For these locals it appeared that DTT was in league with the Establishment in Dar-es-Salaam, so no hope was to be expected from that quarter, but it was still possible to get at one's MP, hence the parliamentary nature of the enquiry.

So the *Modaraba* came to a lingering, rancorous end. In 1991, when the concessions were due to be renewed, the government refused to do so. Nor had it repaid its loan. DTT (for which read Dr Kamel) applied to the International Court of Arbitration in the Hague to have a restraining order placed on the Tanzanian Government to restore its concessions and to pay up.

With Tanzania as defendant, it might be expected that things would happen very slowly. They did; it took three years for the evidence to be provided as requested by the court. In the event, the arbitrators said that the Tanzanians were right to have the concessions withdrawn, but should repay the loan, as and when the foreign currency situation permitted. Perhaps it too still awaits its turn in the morning meetings at the treasury.

．．．．

Having come back through the whole sorry story, and laying aside Tanzania's culpability, could the scheme have worked? The answer is, perhaps surprisingly, yes. Dr Kamel was able to raise U.S.$25 million for gold mining in Tanzania. It was—is—a country with substantial gold resources including a plethora of new discoveries by local people, some of them major. Ten years after DTT's failure, one medium-sized open-cast, heap-leach gold mine is operating and two very large mines are under construction, so at last there seems a chance that the country will fulfill its promise. So where did this well-funded effort go wrong?

The answer has to be that the advice Dr Kamel received was bad. He was led to believe that airborne geophysics can find gold mines, that Tanzania had the potential for numerous, fabulously rich, mines which would allow of high government royalties and that the reef discoveries of the locals should be ignored in favour of a search for virgin eluvial and alluvial resources. No doubt he wished to believe these things himself, no doubt his high-flying style was wasteful of cash, no doubt he was wrong-footed in his dealings with Tanzanian sensibilities, but that does not alter the responsibility.

I carry some of the blame; I recommended pressing ahead with Buhemba which, it turned out, had less gold and was in a much more expensive operating environment than I had expected. It lost money throughout its brief life of under three years.

Chapter Eleven

SUITS AND SACKCLOTH

> *It is complained that some sellers and buyers of shares in mines are fraudulent. I concede it. But can they deceive anyone except a stupid, careless man, unskilled in mining matters? Indeed, a wise and prudent man, skilled in this art, if he doubts the trustworthiness of a seller or buyer, goes at once to the mine that he may for himself examine the vein which has been so greatly praised or disparaged, and may consider whether he will buy or sell the shares or not.*
> —Georgius Agricola, *De Re Metallica*, Book II, 1556.

Because of the tangled and unconventional nature of its dealings, Dar Tadine Tanzania was regarded with deep suspicion by the established mining industry elsewhere. Yet its very failure showed that this enterprise was an honest attempt to go mining. For if you want to make real, big, serious money—say three or four hundred million dollars—and not be in jail for the rest of your natural life, then mining fraud is the place for you.

The arena to do this is in what are the misleadingly named 'penny stocks' of the Vancouver and Toronto stock exchanges. These are midget companies that bump along the bottom in value, occasionally perking up a cent or so when some misguided soul buys a few shares, then slumping back again next day. These shares have no value in themselves, but they do have the vital attribute of being quoted on the exchange.[*]

You buy one of these enterprises of the living dead (not expensive if done quickly; ten thousand dollars will often pick up

---

[*] It appears that for many investors it does not matter what they speculate in, just as long as the stock is a gamble. The low gold price of the late 1990s coincided with a boom in Internet-related stocks, so scores of junior resource companies transformed themselves into investment vehicles for that sector. For example, LatinGold Inc. became Travelbyus.com, Williams Resources Inc. bought Magicorp, a multimedia investment company, and Western Minerals became Adultshop.com, an e-commerce sex-shop.

all the issued shares). Then change the name; make it a good one. Memorable, with a useful acronym something like 'First Imperial Royal Enterprises'. Not some damn fool thing like Bre-X, which you will hear about shortly; you are a running a respectable scam. Now acquire some 'ground'—a mineral concession, somewhere remote with a funny name. Kalimantan? Zimbabwe? Not only far away, but also offering no attractions that might bring a visit from a stockbroker or mining analyst. To ensure that such visitors are rare, mutter about flesh-eating viruses and emphasize the fact that the sanitation at the exploration camp is, well, very primitive.

In this 'exploration play' the geological part is the least of your worries. Ever since diamonds were found in 1991 under a lake in a remote part of Canada's North West Territories by a 'junior' exploration company, there has been money available to put into companies who say they have 'ground prospective for diamonds'. The charm of this statement is that all ground can be prospective for diamonds; you are not lying. Diamonds are found in kimberlite pipes which are ancient volcanic plugs, and these are liable to have burst through any sort of surface strata. From these kimberlites the stones can be washed into rivers and trapped in ancient channels or dispersed into ancient flood plains. All, therefore, 'ground prospective for diamonds', never mind the underlying geology.

Take Zimbabwe. Its only diamond mine was said to have the lowest values in Africa, perhaps the world, and none of the hundreds of thousands of people who have been busy panning for alluvial gold in the rivers there for the past decade have ever reported finding a single stone. Yet Zimbabwe is, or perhaps was by the time you read this, covered with diamond concessions, hundreds of them, granted to all sorts of promoters. Why? Well, diamonds are found in Botswana and South Africa, which are next door, and in Namibia, Angola, the Democratic Republic of the Congo and Tanzania, next door but one, you see. The three golden rules for mineral stock promotions are proximity, proximity, proximity.

Make no mistake, such speculation is at the heart of the business of mining. Without all those doctors and dentists putting some of their wealth at hazard in exploration and mining ventures, then major mines like Sadiola in Mali and Teberebie in Ghana would not exist. Mining finance is all about risk. And reward; in November 1998 the industry giant BHP started Canada's first diamond mine with an output of 3.5 million carats a year—5% of the world's output—worth hundreds of millions of dollars. It arose from that discovery under a lake seven years before, and the owners of the tiny exploration company that achieved it have a free 20% share-holding in the mine.

To appreciate fully the speculative excitement it is worth observing the workings of the market place in mineral rights at the Prospectors and Developers Association's convention in Toronto in early March every year. Here, at the beginning of the Canadian exploration season, to use Mark Twain's terms, the fools meet with the liars. The action is less amongst the sober suits listening to the financial presentations in the conference hall than at the Investors Exchange. Here there are positioned about two hundred 'juniors'—small exploration and development companies— seeking to interest investors large and small. The scene is of a series of booths at which flannel-shirted and corduroy-trousered prospectors and geologists gaze at wet drill cores (wetted to highlight the minerals there), and then use hand lenses to examine them. Dotted amongst these fellows are the sharkskin suits of the promoters, all waving hands and expostulations, proclaiming bonanzas to the cautious-eyed (and more sober-suited) representatives of developers and mining companies moving through them.

March 1997, the conference's 65th anniversary, was perhaps the most noteworthy to date. There were an overwhelming 7,300 attendees (up from the 1996 record of 4,700) and the Trade Show and Investors Exchange had to be moved to a separate convention centre away from the venerable Royal York Hotel that had been big enough to service it in the past.

The Prospector of the Year title was given to one John Felderhof, a Canadian of Dutch origin and with a short fuse. He was the vice chairman and head of exploration of an exploration outfit peculiarly called Bre-X. This had achieved the nirvana of every junior; it had evidently found a true El Dorado in eastern

Kalimantan, the Indonesian part of Borneo. Seventy-one million ounces of gold, or about 2,400 tonnes, worth about 25 billion dollars, more than a year's production from the entire globe. Little wonder that Mr Felderhof effervesced around the crowded halls, acknowledging plaudits like a man possessed.

Proximity had done the trick. In the early 1980s an old geological phrase became a hot new one. Epithermal gold. Although a number of mines in the United States and elsewhere were thought to be of this sort, and one of these in Nevada, known as the Carlin trend, was shaping up to be North America's biggest gold source, big new gold deposits were also being found on the western rim of the Pacific that did not fit into the conventional granite source rock pattern. This region, an arc stretching from Japan to New Zealand, is where 80% of the world's volcanoes are found and the promoters gave it a hot new name as well, the Rim of Fire.

In particular there is an area where four tectonic plates meet, stretching from Papua New Guinea (PNG) to the Philippines and northern Sumatra, which is fairly hotching with volcanism. A number of discoveries had been made and some big mines— Ertsberg (gold), OK Tedi (gold), Bougainville (copper and gold)— discovered. Yet it was not until 1987 or so that the way in which plate tectonics creates epithermals became public property, so as to speak, amongst geologists and promoters.

Epithermal gold is a younger variation of the hydrothermal sort. However, it has been picked up from its original source rocks by rainwater, which has drained down and been returned to the surface by volcanic heat, often appearing as hot springs. In the process—and some active geysers today have been going for up to a million years—this water dissolves gold and other metals at depth and reprecipitates them nearer the surface.

Because of this the gold (or, very often, copper and gold) frequently appears as a shallow, disseminated ore body in the vicinity of the volcanic source providing the heat. So typically an epithermal orebody is ideal for simple open pit mining. More, because it is usually in a well-weathered tropical environment, often all that is needed to get the gold out is to pile up the ore in heaps and run cyanide solution over them. On the other hand it is also frequently a 'blind' deposit, lying hidden below the site of the original hot spring.

This theory was briskly retrofitted to a number of big existing mines in Australia and the Americas, and it also became clear that there was the potential for epithermal gold in previously unconsidered parts of the world with hot springs (Turkey, Greece, Italy, Mongolia, Cornwall ...).

One of the important features of this type of deposit is that the circulating fluids alter the surrounding rocks, causing them to contain zones of clay (remember Buckreef's problems from surface water draining down?) and hydrated silica, the latter showing up as semi-precious stones like chalcedony and jasper. Remember you read it here first; I was there, in the Turkana desert of northern Kenya, close to the Rift Valley, when Göran Petersson, the geologist I was working with, looked at the gold the local tribeswomen were winnowing from the dusty ground, ferreted amongst the stony gravels they were working on and said, 'This looks like an alteration ...' You read it here first, and probably last. That's mining for you.

Once epithermal gold had been identified as a feature of the vast 'Rim of Fire', there was a brisk move by stock promoters into the area, and any mineralization on their properties was forthwith identified along the lines of 'strongly suggestive of epithermal activity'. Bre-X was ostensibly the biggest such discovery and in its full flowering was described by its promoters (in early 1996) as 'a seven kilometre- long epithermal gold system spatially related to a large copper/gold porphyry. It is now believed that it has a minimum resource of some 30 Moz (million ounces) of gold'.[53] Bre-X shares had been as low as four Canadian cents apiece in 1992; by now they were worth over ninety dollars, and the shares began trading on the Toronto stock exchange. (Up to that time they could only be bought and sold in the tiny Alberta exchange.) Once on the big board, the shares went to C.$230 apiece, bringing the company's capitalization to four billion Canadian dollars. To contain the numbers to manageable amounts the shares were split 10:1.

This was a true feeding frenzy. It was pointed out at the time that this price gave the company a valuation per ounce of gold in the ground of over twice that of the largest gold mine in the world. This was Anglo American's Western Areas in South Africa, which had 50% more gold to boast of at a much higher level of certainty, was fully funded and was (and is) in production.[54]

No matter. Bre-X was the hottest stock to hit the Toronto exchange for fifty years. It was launched at a fortuitous time, for geologists from one of the many Canadian diamond exploration stocks that had proliferated since that unique discovery under a northern lake had just stumbled, almost literally, on perhaps the largest deposit of nickel in the world while fossicking for gems in Labrador. The market was in a state of hyperactivity; the brokers were eager to find another big-bonanza stock, the analysts eager to identify it, the fund managers eager to get hold of it. Perhaps if these enthusiasts had known how Bre-X had been put together they may have been less sanguine about its prospects.

The Bre-X story began in 1987, when a stock promoter called David Walsh incorporated Bre-X, with himself and another company of his, Bresea Resources Ltd, as the principal stockholders. By 1993 Walsh was personally bankrupt, although not in a particularly grand way; he had C.$71,000 in debt on no less than seventeen credit cards. Interestingly, though, he was able to scratch together a few thousand dollars (his wife was believed to have supplied it) to go to Indonesia where he met John Felderhof.

In Indonesia a group led by an Australian company, Westralian Resource Projects Ltd, had used its gold exploration subsidiary, Montague Gold, to drill twenty holes to sixty metres depth (about the limit of a big open cast mine) on a gold prospect near the upper reaches of the Busang River, in eastern Kalimantan. Trenching had already shown some gold, not much, about three grams a tonne at best. Montague's work showed a that the gold potential was doubtful; there could be a big, low grade resource, maybe twenty million tonnes at two grams a tonne, but the emphasis was on the 'could be'. Gold values were low and erratic. Still, a could-be figure of over a million ounces (thirty-one million grams) in a remote jungle setting was something an astute promoter might make something of.

Improbably, Montague was acquired by an Edinburgh mining finance company in 1991 (which had also invested in a loss-making Scottish coal mine) and it is here that John Felderhof enters the story. In 1986 he had worked in the Busang area, following which he operated out of Jakarta. He was, therefore, a natural choice for the Scots, who now controlled Montague, to review the Westralian figures. They asked him to put forward a plan and a budget to follow up these mediocre results, which he did. More, as with every good geologist that I have met, he had a convincing theory, stiff with impenetrable terminology, as to why there should be untold riches (of gold in this case) in one particular place. His idea, which was also held by his Filipino sidekick, Michael de Guzman, was that there was a geological feature known as a maar diatreme, in which volcanic activity caused pockets of high mineralization at the vents. It will be noted that this opaque phrase was not much applied when publicly describing the deposit; here the key code words *epithermal* and *Rim of Fire* were used, and the names of successful mines on similar geology were bandied about.

De Guzman, perhaps not incidentally, was, it turned out, a man with at least four wives, one in Manila and the rest scattered around Indonesia. A Deloitte Touche report on the scandal—the Bre-X scandal, that is—said that 'most' of these ladies were unaware of the existence of the others.[55] De Guzman had actually been fired from his last job for the misuse of funds—a relatively small amount, of the order of $1,000—that had been used covertly to support one of these. He also had, it was later said, a tendency to hype his geological findings.

In any event, armed with their theory, Felderhof and de Guzman spent four days in the Busang concession, sampled around the drill holes and suddenly found a possible additional underground resource of 60 million tonnes at 3.5 g/t.*

---

* People in mining are careful to distinguish between 'resources' and 'reserves'. There is a whole hierarchy of categories of certainty of what ore is in the ground—measured, indicated and so on. However, a resource is only what may be there, a reserve is something you are sure of. Right up to the end almost all of the Bre-X find was in the 'resource' category. It did not appear to worry shareholders.

All poor together

At the time Felderhof was chronically short of money; he, like Walsh, had suffered from the collapse of the stock markets in 1987 and had sold what was described as his 'mansion' in Perth, Western Australia, and set up in Indonesia, in an attempt to survive as a freelance consultant. By now the risks endemic to such a business will be understood by the reader.

So here we have a slightly gamy Filipino field geologist, a Canadian-Dutch consulting geologist with a great theory and a shortage of cash, and a bankrupt promoter with effectively nothing more than a knowledge of how to work the stock exchanges.

Through the sort of complex dealing that typifies junior exploration companies who have little to trade but a lot of shareholders to impress, Montague gave Walsh's Bre-X a purchase option on the property for a very small sum of money—$80,000. Just how Bre-X found that amount is not sure, but they did. Indeed a total of $200,000 was scraped together, enough to put a modest exploration team in the field.

However, once in place the resource figures began to leap skywards, together with the Bre-X stock. Then Montague were found to be surprisingly difficult to offload. There was a bitter renegotiation of the deal and its fine print was discovered to require that Bre-X shell out U.S.$1.7 million for the property legally to change hands.

By that time it did not matter much; such money was small change to Bre-X stockholders. The bandwagon was launched in April 1993, with a report to shareholders. This did not confuse them with geology much beyond the Rim of Fire thing, but it did say that one million ounces of gold could be recovered from the deposit at a cash cost (i.e., excluding capital recovery) of $155/ounce. At the time the gold price was around $350 an ounce.

In 1994 the Bre-X drill results in the same zone were showing grades of twice to three times that amount, and a year later drilling was down to 240 metres with grades of 10–15 grams a tonne. The original mineralized area of one kilometre on strike and 200 to 300 metres on width had grown to 3.5 kilometres and up to 500 metres wide. A respectable Canadian company, for whom it would be a kindness not to be mentioned yet again in this connection, undertook a reserve calculation on the 21 kilometres of drill core results they were provided with that gave a grand total of 39 million ounces. This is more gold than Zimbabwe produced in the whole of the second half of the twentieth century. No doubts were expressed; the tests themselves were being undertaken by a company that was, and still is, a reputable assayer.

David Walsh was now speaking of a 50-million ounce resource, and was asked to give the prestigious final presentation at the 1996 Prospectors and Developers Annual Convention. It was packed solid with awed suits. 'World Class' was on everybody's lips. There were 160 applications for concessions in Kalimantan piled up on the desk of the Indonesian Government's Director General of Mines.

As the purported scale of Bre-X grew, so did a ring of counter-claimants about it. First of all, there were the Indonesians. Twenty percent of the shares were already held by 'local interests'. From my own experience with a small coal mine in that country, the army, or rather the local general, would have had some sort of holding in those. However, now up pops another local interest, a Mr Merukh, who laid claim to a 40% share-holding. This arose out of some arrangements made in those early fraught days when it was necessary to get Busang away from Montague as quickly and as discreetly as possible. Mr Merukh threatened a U.S.$1.9 billion lawsuit.

The situation was complicated by the fact that the Bre-X Busang properties were in three areas, and some of the local interests found out that their holdings encompassed only one of these. An Australian company, Golden Valley Mines, also laid claim to the deposit, forming an alliance with Mr Merukh. The managing director of Golden Valley Mines turned out to be a Mr Warren Beckwith, who had been MD of Montague Gold NL at the time when Bre-X obtained its purchase option. Bre-X described the claim as 'false and misleading'. But in October 1996 the Director General of Mines withheld permission for further exploration by Bre-X until the dispute was settled.

Besieged, Bre-X needed powerful friends. It found them. It entered into what it correctly described as a 'strategic alliance' with the Indonesian company PT Panutan Duta. This was, not incidentally, controlled by the then President Suharto's eldest son, Sigit Harjojudanto. In return for 'consulting services', PT Panutan Duta would receive a million dollars a month and a 10% free interest in the richer two of the three Busang areas. Thereafter, little more was heard of competing interests and Bre-X shares went up by 16% on the news. It seems a little unfair, if not superfluous, that the Director General of Mines was fired as well.

So that was settled. The next problem was to find someone to buy the discovery. Remember that Bre-X was an exploration company and that in the usual way of things, as the previous chapter explained, it would have sold out to a wealthier developer who would finance the proving up of the deposit, with Bre-X retaining a small shareholding as part of the deal.

But for a leviathan like Busang it was not necessary to find an intermediary who could fill out the gaps in the geological story and put together a 'bankable document' to attract mining companies; the miners were already banging on the door, braying to be let in.

At one stage or another in latter half of 1996, every major gold mining company in the world was said to be holding discussions, or be 'in negotiation' with Bre-X: the South African majors headed by Anglo American, Rio Tinto of the United Kingdom, Freeport McMoran of the United States, Teck of Canada and so on. However, by early 1997 it had settled down to a race between the two top Canadian gold miners, Placer Dome and Barrick Gold Corp.

At first Barrick, North America's largest gold producer, seemed to have the acquisition sewn up. In November 1996 it confirmed that it had formed an association with an Indonesian construction company controlled by President Suharto's eldest daughter. A Mr Vincent Borg, who as Barrick's vice president of public affairs had the uncomfortable task of addressing a sceptical press on the decision, said that it was a necessary step 'to prepare for a potential mining development project ... you need local suppliers. You need roads built ...' He declined to say whether or not Barrick intended to make an offer for the Busang discovery. Two weeks later he came clean; they were 'in negotiation' with Bre-X. These negotiations were, according to Bre-X, under the 'guidance' of the government, which had also said that 'it would appreciate' if the parties could consider giving it a 10% participation. From this it was inferred that Barrick was the miner of (government) choice, so it was a done deal. Barrick shares went up by nearly 8%, adding almost a billion U.S. dollars to their value.

Then, unexpectedly, in January 1997 Placer Dome made its play. The company side-stepped the murky (and expensive) business of acquiring friends in court in Indonesia by making a direct bid for the Bre-X shares, promising a new joint company, a proposed investment of U.S.$ 1.7 billion in a mine and up to 40% offered to Indonesian investors. Amazingly, although Placer Dome produced about 1.5 million profitable ounces of gold in 1996 and Bre-X produced none, at that time the two companies' market capitalization was roughly the same at around U.S.$5 billion. David Walsh, a man who stood to benefit hugely in the long term from such an arrangement (shares in Bre-X rose 10% at the announcement; they had fallen on word of the Barrick deal) said the proposal 'cast a whole new light on the project'.

There seemed a high probability now that the Placer Dome deal would go through, as being the most favourable to the Bre-X shareholders. Control would stay firmly in Canada and 'local interests' could be relegated to bidding for the 40% of shares assigned to Indonesians like anyone else.

This would not do.

Presidents of sovereign states are supposed to be above the rough and tumble of the market place; to look after their financial affairs they will have a 'man of business'. For what it is worth, the three I have known were uniformly self-effacing, physically undistinguished and quite ruthless.

The now ex-President Suharto's man of business was, perhaps even still is, Mr Muhammad 'Bob' Hasan. In this crisis he acted with, for Indonesia, breathtaking speed. First, in a matter of days, he acquired command of the Indonesian companies already holding an undisputed 10% interest in Busang. Then, a couple of weeks later, at the end of January 1997, he used a company controlled by a foundation owned jointly by himself, the

All poor together

president and the president's eldest son (Sigit Harjojudanto, already getting a million a month from Bre-X) to buy up the existing 'local interest' component of the giant Ertzberg/Grasberg copper-gold mine in Irian Jaya. In gold terms alone this is one of the world's great mines. Operational since the middle 'sixties, it now produces over a million and a half ounces a year. The company that runs it is the New Orleans-based Freeport McMoran Copper and Gold.

In the middle of February 1997, less than two weeks after 'Bob' Hasan had consolidated the presidential interest in Grasberg, Bre-X announced that Freeport McMoran would take a 15% interest in the Busang project, would provide funding totalling U.S.$1.6 billion (Chase Manhattan Bank provided a U.S.$1.2 billion line of credit) and would be the sole operator. Bre-X would retain 45%, 30% would be held by the 'local interests' companies, now ultimately controlled by the president and 10% by the government, also, of course, ultimately controlled by the president. Adding together all the bits and pieces, Busang would pretty well be a Suharto family business. So that wrapped up who the miner was to be and, for that matter, who controlled Busang.

David Walsh put a brave face on it, saying that 'the Busang deposit is of national importance to the Republic of Indonesia.' He added some extra spice by announcing at the same time that its consultant's latest calculations showed that the total gold (in all categories of certainty) was over seventy million ounces. The new joint venture would be called Busang Indonesian Gold, giving it an appropriate acronym.

· · · ·

Apart from its funny name, I had not taken much notice of the Bre-X excitement. There seemed no reason to disbelieve the tales of great riches; I had once been down an Indonesian gold mine that was hoisting ore of nearly an ounce a tonne, five or six times as rich as the average mine in Zimbabwe. Indonesia was an exciting place for gold discoveries, and it was possible that Bre-X's suggestion that it had made the gold find of the century was not entirely promotional hype. But then, in March 1997, I read about a visit to the exploration camp in the Canadian mining newspaper, *The Northern Miner*.

There are only three or four major mining magazines, all with lengthy antecedents, modest premises and minimal staff. The weekly *Mining Journal* of London started in 1835, and was the world's first industrial newspaper. From the same stable comes the *Mining Magazine*, which started in 1909. Its principal instigator was Herbert Hoover, a widely-travelled mining engineer who was later to be the president of the United States. In America it is the *Engineering and Mining Journal*, which was started in 1866 and is based in Chicago. The *Northern Miner* was founded in 1915 at Cobalt in northern Ontario. At the time the policy of Canadian mining publications was that only the good news be printed. A motto of 'On the Level' was adopted by the new publication; in the Bre-X fiasco this credo was given—and survived—a severe test.

The article in the *Northern Miner* was about a video taken by a mining analyst who had visited the Bre-X camp in the jungles of Kalimantan. There had been some stories floating around to the effect that the assay results were unusually erratic and there were no drill cores. In fact there were some bits and pieces of drill cores, as the video showed. However, the rest had been used up, said Bre-X, by assaying and testing. This is unusual; the standard practice is to split cores lengthways, with a half or, at a minimum, a quarter being kept back against disputes or queries. Indeed, a site like this is usually dominated by piles of long, shallow core-boxes, each well labelled, and with the half-cores nestling in compartments inside, carefully marked with depths and assays. Unusually too, for an exploration camp, there was a metallurgist on site to look after the sampling. Like the rest of the geological team he was a Filipino; his name was given as Jerome Alo.

We have met, you remember, Jerry Alo in the previous chapter. He had arrived shortly after I had started up Dr Kamel's Buhemba plant, and had been there at the time of the first clean-up—the great smelting disaster that culminated in my handing over two pathetic little pieces of gold at Jomo Kenyatta Airport to Dr Kamel. Jerry's job had been to take over the day-to-day control of

All poor together

the operation; he had been hard-working and cheerful like most Filipinos, and we had shared a tent in the primitive early days while he picked up the ropes. Eventually he took over the running of the plant and I faded out of the picture.

Some months later I heard indirectly that he had left. Did that have anything to do with the embarrassingly poor recovery from the first clean-up? It seemed unlikely, and anyway it was old, mortifying, history. In addition, Dar Tadine Tanzania was gradually fading away in a string of protracted legal actions, and my own contribution was long over. But here was Jerry again, improbably cast as a member of an *exploration* team, a team that was having some credibility problems. I started to follow the unfolding Bre-X story more closely.

Just a week after the *Northern Miner* article there came a story from an Indonesian newspaper that it was all a fake. 'Nonsense,' said David Walsh, the chief executive of Bre-X. A press announcement said that 'the continuing proliferation of falsehoods and misinformation based on unsubstantiated allegations by unnamed sources ... has prompted Bre-X to consider legal action against certain parties and publications ...'

Coincidentally, the *Mining Journal* issue that carried this combination of denials and threats also had an article about my own efforts to encourage mining investment in Zimbabwe (and perhaps to use the mining consultants therein). In it I told of how our several hundred mines had painstakingly raised our annual production of gold from 14 tonnes to 25 tonnes (800,000 ounces) in the past fifteen years. Quite a lot of this splendid growth—all of about six tonnes a year (200,000 ounces)—had come, gosh, from foreign venture capital. Predictably, investors were underwhelmed by this tiny, sluggish bandwagon when there was a very much bigger and glitzier one; about 300 times the size, in fact.

•   •   •   •

The problem with a successful scam is to know when to end it. It is necessary to have an exit scheme that allows the perpetrator either to vanish utterly—a suicide note is just the thing—or to appear as confounded as everybody else, but sadder and wiser, bearing losses and recriminations with dignity, and with all the loot well hidden.[56]

Very well in theory, but for the Bre-X fraud perpetrators their nerves must have already been stretched to breaking point by the time the Freeport deal was abruptly imposed on the company. They had come perilously close to being exposed when the nascent agreement with Barrick resulted in the start of due diligence work at Busang by that company. Luckily Barrick did not sink any check boreholes, but it did run assays on such limited drill core material as was available. To their surprise there was very little gold in them; only 5 out of 135 samples gave results above one gram a tonne.

Both Barrick and the Canadian office of Bre-X gravely shook their heads over these 'anomalous' results (there was some talk of poor assay practice), and Barrick collected a fresh set of samples. These gave the expected figures, and presumably both parties heaved sighs of relief, although perhaps for different reasons. The only other oddity was that the Canadian laboratory undertaking the assays found that the gold in the samples was unusually coarse. But by then Barrick was out of the loop.

Barrick may not have drilled, but de Guzman was severely rattled. He threatened that if they did, 'I will pack up as soon as drilling starts. Same response by our senior field staff... Regardless of the outcome'.

They had been lucky so far. But at some time in the future a full due diligence would be undertaken by whoever had finally bought Busang, and the fat would be in the fire. That might only be a few weeks away now; it was time to cover their tracks and cash in.

On the night of 23 January 1997 a fire destroyed the main geological offices at the exploration camp. In the process it consumed the information on the location of the drill holes as well as the logs of the drill cores. Thanks to the efforts of three Canadian geologists who happened to be there at the time, the core shed and its contents (such as they were) was saved. Shortly after daylight, Jerry Alo, who was in charge of the camp, ordered the charred wreckage of the buildings to be bulldozed into the ground; he gave no opportunity to check for the origin of the fire or to recover any data that had survived. Later he said that he thought the cause was an electrical fault.

De Guzman, who was away at the time, improbably described the event as a 'normal—minor incident' in a memo to Felderhof and said that nothing had been lost that was not duplicated elsewhere. However, when eventually a full due diligence did get under way a few weeks later, critical data on the location of drill holes was missing, and the memories of those who had originally surveyed their positions had to be drawn on to find the sites.

Now to cash in. Up to that point Jerry Alo had bought and sold Bre-X shares for a profit of about U.S.$ 1.5 million. De Guzman's project manager at the site, another Filipino called Caesar Puspos, also deeply implicated in the scam, had made about U.S.$1.7 million. By the time an accounting was made of what de Guzman had got out of it he was dead, but it was thought that about U.S.$4 million had been put into the safekeeping of his various wives and girlfriends.

But then came the bombshell of the abrupt announcement of the Freeport deal; it appears to have come right out of the blue to the workers in the field. Less than two days later, de Guzman faxed a memorandum to John Felderhof, asking him to arrange the sale of the remaining shares on which they had options. For each of them profits of the order of ten million dollars or more were at stake.

There appears to have been no immediate answer. However, almost the entire exploration team had been invited to the Prospectors and Developers Annual Convention (PDAC) about two weeks later, where John Felderhof was to be named 'Prospector of the Year'. A fellow-passenger noted that it was a happy group of Filipinos, chattering in Tagalog, who were sitting in business class on the long flight to Toronto.

But there, to their horror, Felderhof refused to let them sell their shares; it was, he informed them loftily, 'not an appropriate time' as the Freeport due diligence had already started. (They may not have known this, having been away from the site for some days, but the first of four Freeport holes, each adjacent to a Bre-X one, had already been completed when they saw him.) Felderhof could afford to take a high ethical stand; he had made nearly U.S.$30 million when he quietly sold his own shares between April and September the previous year. The Filipinos were being dumped.

Felderhof was not the only one who had taken profits. Insider trading information from the Ontario Securities Commission, the regulators of the Toronto stock exchange, showed that David Walsh and his wife Jeanette (who was the Bre-X corporate secretary), together with Felderhof, had made a combined profit of about U.S.$60 million the previous year.

The end was very near now. It had not been necessary for Freeport to wait for the drill-hole results; they found some complete Bre-X core that had not been used up in the unusual sample preparation system practised by Jerry Alo. It had no gold in it.

On the last day of the PDAC meeting in Toronto, on Wednesday 12 March 1997, Jim Bob Moffett, chairman and chief executive of Freeport McMoran phoned David Walsh at his hotel there with the news. Walsh had Felderhof recalled from his manic progress around the halls and asked him to call Moffett back. Felderhof did so, then phoned Walsh to tell him that he was going to fire de Guzman—but first he wanted the geologist to go back to Busang to 'handle this issue'. Walsh, bewildered, asked Felderhof if he was going there as well, but the latter replied that it was enough that de Guzman was. Felderhof's assessment was that Freeport's 'assaying is wrong'.

Then there followed, unknown it seems to Walsh, a four-hour meeting in Felderhof's room with de Guzman and Caesar Puspos. The next day all the Filipinos left for Hong Kong, where they would split up, de Guzman being the only member of the team to return directly to Busang. In contrast with the flight out, an Indonesian businessman who was a fellow-passenger and who spoke with de Guzman, remembers that the group was upset because they had been unable to cash in their share options.

On 19 March 1997, de Guzman climbed into a small helicopter,

All poor together

an Alouette III, to fly from Samarinda on the coast up to the site. He was on his way to meet the Freeport due diligence team, who by this stage had accumulated sufficient information to know that the whole saga was a massive fraud.

Three passengers can sit on the bench at the back of an Alouette; de Guzman was the only one. Along with many others I have travelled in these lightweights; they were the principal helicopter of the Rhodesian Air Force. Then we flew with the sliding doors locked open anyway, clutching our weapons and hoping desperately that the worn webbing lap straps were up to the job of holding us in as we tipped crazily over the bush.

When closed the doors are moved by simply pulling back on the handle, and twenty minutes into the flight it was easy enough for de Guzman to drag the left one open and walk out. 'Suddenly, pop, I heard a loud bang,' reported the pilot. They were travelling at about 130 mph at the time, 800 feet up, over deep, swampy jungle. De Guzman may not have died immediately on impact, but he was very dead when found, floating in waist-deep water, four days later. In the helicopter he left behind a bag containing his cellular phone, watch and other minor valuables plus a number of farewell notes; to Felderhof, to various colleagues and to the wife and her children who lived in the Philippines.

* * * *

Even before the body was found the truth had started to leak out. For the Freeport team at the camp, the last pieces of the puzzle had fallen into place on the very day de Guzman leapt to his death. When the news burst upon them, they were hastily evacuated to Samarinda, whence they were flown by the company jet to Jakarta, registered in hotels under assumed names, and told to talk to nobody.

Two days later came the Indonesian newspaper report that provoked such vigorous denials and threats from David Walsh. He had reason; it caused Bre-X stock to lose C.$800 million. But it was only the first rattle of stones of the avalanche. On 25 March 1997 the persistent rumours led the Toronto Stock Exchange to suspend trading in the Bre-X shares.

Up to that point the Bre-X executive had been publicly denying every accusation. But by the next day the Freeport evidence was so overwhelming (it included the discovery of nearly 4,000 'duplicate' Busang assays whose results showed negligible gold compared with the originals) that they had to admit their own fears. They confessed that a week before they had asked a consulting company, Strathcona Mineral Services, to audit the exploration operation.

At that admission the shares, which had been trading in Toronto the previous week at over C.$20, fell to a tenth of their value, and the computerized system of the exchange collapsed under the weight of 'sell' orders. On 31 March the *Northern Miner* (which had, awkwardly, already named Walsh and Felderhof as its 'Mining Men of the Year') had a headline saying: Freeport drilling returns 'insignificant values'. Not unrelatedly, the Toronto Stock Exchange trading system crashed again that day.

The rest can be quickly told. Felderhof had grudgingly cooperated with the investigators, coming to Busang at the end of March for de Guzman's funeral, but on 3 of May he was safely ensconced in the Cayman Islands, from whence he announced that he was '110 per cent confident the gold is there.' Two days later the Strathcona independent audit was published. The authors said bluntly that Bre-X was a fraud 'of a magnitude and a precision that, to our knowledge, is without precedent in the history of mining anywhere in the world.'

Next day, with the shares at nine cents, the stock was delisted and the police moved in, the Mounties in Canada and the Indonesian police at Busang. Two days later, on 8 May, Bre-X filed for bankruptcy while its directors made a rush for the exits, denying everything as they went.

So, to wrap the story up, obviously the samples sent for the cyanidation assay were salted; that is, intentionally contaminated with gold from elsewhere. But when, where, how and by whom?

The investigators believed that salting started as far back as late 1993. That was when the option was taken up and the Bre-X promotional bandwagon was first run out. The reality was that the results from the first two holes that de Guzman drilled were as bad as those drilled four years before by Westralian. It was too late to do anything about them, but the cores from the next two were crudely salted, giving some almost embarrassingly high results. Yet it did the trick; the figures ramped the Bre-X share price up enough to gain funding for further drill holes—eventually about 140 of them if the numbers can be believed.

But the figures from the first attempt were unsatisfactory if a protracted fraud was to be achieved. The assays were all over the place, ounce-a-tonne material alternating with barren portions. The impression was of an erratic reef structure, not the steady and extensive values required to give the impression of a major discovery. This was because a combination of slivers of gold from jewellery and pure alluvial gold had been used for the salting.

It was necessary to get something more professional arranged for plausibility to be sustained. Jerry Alo came on board in early 1994 and thereafter the figures steadied down.

However, he, and Caesar Puspos (who probably did most of the actual salting), were not quite sophisticated enough in their efforts; once suspicion was aroused the salting was quickly unmasked. Again, the clue was the nature of the gold particles. Not only were they much coarser than they should have been (often of a size to have been visible in the drill cores, something nobody, Indonesian, Filipino or Canadian, had ever reported seeing), but they were of the wrong shape. Under an optical microscope they were rounded, as though they had been rolled and abraded. Under an electron microprobe—an electron microscope that analyses the object at the same time—the silver content was found to be far greater than that of such gold as had been actually found in untouched Busang samples. However, this silver was almost all in the central cores of the particles, while the outer parts were largely barren. They were, without a shadow of doubt, grains of alluvial gold; silver tends to leach out (very slowly) in water, while gold, of course, is almost impervious.

If further confirmation was required, the samples contained quantities of the heavy minerals usually associated with alluvial gold, such as rutile and magnetite. It would have been most unlikely for these to have been present in those amounts in the original cores.

The discovery of the associated minerals also showed how the salting was done. To accurately measure out the precise amount of gold required is very difficult. The two-metre length of drill core used by Bre-X (less the inadequate 20-cm piece they kept back for 'checks') would have weighed about ten kilograms. For the average gold grade of 2.5 grams a tonne that we may assume the fraudsters were aiming at, the weight of the gold needed would have been about twenty-five thousandths of a gram. Because of the high density of gold, this would have been represented by a barely visible pinhead, yet that in turn would have to be split into scores of sub-particles to simulate the distribution of gold that a reasonably steady grade would require. Even with fine alluvial gold this would have been a near-impossible task. Indeed, the feature known as 'the nugget effect' has always bedevilled gold assays; it only requires an infinitesimal amount of the metal to send a result soaring to completely misleading levels.

So the degree of manipulation of very fine gold particles is apparently daunting. What is needed is not the gold itself, but what is actually easier to get, a concentrate with fine gold dispersed through it. My own experience of salted mines has included a case in Zambia, where a pyrite flotation concentrate containing a few grams a tonne of gold was used to perk up a very low grade tailings dump that the holder was trying to sell. The deception was completely successful; it was only when the expensive plant that the new owner had installed was started up that the horrid truth emerged.

So de Guzman and company must have taken concentrates panned from the Busang streams by the local Dayaks (significantly there was very little such gold). They probably repanned them, kept for themselves the high-grade 'tail' and salted with the lower grade mass of heavy minerals at its head. Here the gold would have been finer anyway. The technique requires no micro balances or clever manipulation. The amount added does not even need to be weighed out; once a reasonable idea of the volume needed has been obtained, any small container would do to measure it.

All poor together

This was still not enough to minimize the nugget effect. They needed to use an assaying technique that would require a large bulk of material for each determination, so giving a reasonable chance that the sample being tested had a representative amount of gold in it. So instead of using fire assay, which has for centuries been the only reliable way of finding the total gold content of a sample and which typically requires a sample weight of 50 grams or less, they instructed the laboratory they employed to assay by cyanide extraction. This uses a kilogram or so, and has the merit that not only were the chances of erratic results from the relatively coarse particles used for salting correspondingly reduced, but there was a credible explanation for the absence of spare material for checking.

The actual operation seems to have been carried out in Busang at first, where in 1995 a sample preparation laboratory was built. However, it worked out easier for Jerry Alo to oversee the crushing of the cores as part of his general camp responsibility, to split them down ready for the cyaniding test and to send them on to Samarinda where Caesar Puspos intercepted the sample bags and salted them. Remarkably, almost everybody who ate at the company mess in Samarinda reported seeing numbers of reopened sample bags.

Frustratingly, David Walsh died of a brain aneurysm in June 1998, while staying with his wife at his beach front house in Bermuda. (The liquidators, Deloitte Touche, allowed them a mere U.S.$3,000 a week to live on.) John Felderhof, the sole survivor from the principal suspects, must be feeling uncomfortable; it is all very well to stay in the Cayman Islands, but even there the authorities don't want too many high-level fugitives hanging around indefinitely. From the experience of others it may be assumed that the fortune he made from Bre-X is being drained away by payments to retain his residential rights, and that at some stage he will find this intolerable and will break cover.

Inevitably, rumours abounded that de Guzman had been pushed, not fallen, from the helicopter, or that the body in the jungle was not his. But it seems unlikely; on that day only the Freeport team knew the whole story of the fraud, and they were at the camp,

Suits and sackcloth

sworn to silence. Disbelief was the first reaction from everybody else. That included the Indonesian military, whose involvement in the economic life of the country must have included big chunks of the Bre-X portfolio, and who would have been prime suspects for any murder.

It is manifest that Michael de Guzman, Ceasar Puspos and Jerry Alo were involved. There are also suspicions over three or four others—Caesar Puspos' brother who was also at the site, and his girlfriend who worked for the company, and so on. But who else knew?

This is a most important question. Of course, once the story broke there was the usual barrage of 'told-you-so's' and some sniffy comments from the Australians about the lack of reporting standards on the stock exchanges in Canada comparable to those used in Australia.[57] In particular there was condemnation of the way in which the terms 'reserves' and 'resources' were used indiscriminately by the participants, although they have strict legal definitions. But although a 'Mining Standards Task Force' was set up in Toronto to plug the gaps in the system, it can be safely predicted that neither Australia nor Canada will necessarily be immune to such scams in the future.

Look at it from this point of view. Bre-X was a company that had gold assets in the ground valued by its shareholders in the range of several billions of dollars; one authoritative estimate had it that they represented about 8% of the global total of gold reserves and resources.[58] These assets were based solely on drill core assays. Yet the whole sampling procedure on these cores was half-baked, not to say fundamentally flawed.

Ever since gold has been won there have been swindles based on falsified evidence. Because of this the mining industry has evolved elaborate rules and guidelines to enable companies to protect and reassure investors. It is not only geologists who know of such matters; stockbrokers, analysts and fund managers, even promoters like David Walsh, are aware of them as well.

Yet from the start Bre-X broke rule after rule. The cores were not split, independent verification was not used, sample bags were re-opened before arrival at the assay laboratories. Why were red flags not raised at an early stage? Felderhof, de Guzman, Puspos and Alo were clearly implicated, but there were at times upwards of a score of geologists and other mining professionals at the exploration camp and in the offices in Samarinda, Jakarta and Toronto. These were people who knew all about the need to retain credibility in such situations. Was *everybody* in on the fraud? As the *Mining Journal* said in an incredulous editorial, 'Thousands of samples have been collected and assayed over the past two years and it is difficult to believe that the salting of samples on such a scale could be feasible ...'.

The answer has to be that at the back of everybody's mind there were suspicions, but that nobody wanted to blow the whistle. (There were a couple of honourable exceptions, and one of these may have lost his post as a result. In any event, no notice was taken of their views.) Jobs in the exploration business are fickle; for example, in 1999 the slump in the gold price resulted in an estimated 1,800 Australian geologists becoming unemployed.[59] Neville Shute had some pertinent observations to make on this type of situation, relating to his experience in trying to fund naval research during World War II:

> These naval officers were as brave as lions (at sea) ... but to be asked to risk their jobs on a verbal decision involving public money often seemed to them unfair.
>
> Now and again, we would find some cheerful young commander or captain who was not affected by these scruples, who was as brave in the office as he was at sea. Commenting on such a regular officer and his way of doing business we would say, 'He's a good one. I bet he's got private means.' Invariably investigation proved we were right.[60]

He went on to bemoan the effect of taxation and death duties on the quality of administration; diminished personal wealth was undermining the courage of those in positions of responsibility. Nowadays the best hope of gaining such wealth while employed is through stock options in a growing company. But this has a contrary effect to real financial independence, as it further discourages speaking out. That, I believe, was what happened with the professional staff at Bre-X. All the management had immensely profitable stock options, many of whom exercised them, while all the professional staff below that level aspired to a promotion that would enable them to share in the bonanza. The Bre-X fraud was only exposed when outside investigators became involved. Until then it is my belief that the tight inner ring of conspirators comprising Felderhof and five or six of the Filipino field staff was protected by a broad, and probably largely unspoken, conspiracy of silence.

Chapter Twelve

RECKLESS ASSERTIONS

Statements such as these, betraying weak powers of observation, strong fancy, an eager craving for wonders, and childish reasoning, could not fail to awaken distrust by their intrinsic demerits ...
—Desborough Cooley, Royal Geographic Society, London, 1848, on hearing a report by a German missionary of snow on Mount Kilimanjaro

The development assistance industry is, of course, a lot more dignified than the hurly-burly of exploration plays in the stock market. It is dominated by big, staid institutions, occupying massive fortresses close to the White House (the World Bank and IMF), or handsome headquarters in Geneva (the World Health Organization, WHO) and Vienna (the United Nations Industrial Development Organization, UNIDO). Nor are these closed bodies. The World Bank in particular spends about one percent of its budget annually on 'External Relations' and has an impressive system of peer review for the internal and external reports that it issues. If anything these agencies devote too much time to agonizing over how to become more effective, trying to incorporate all their burgeoning concerns (the environment, women, poverty alleviation, capacity building, governance, HIV/AIDS—each year seems to bring a new one) into every project.

The openness of the World Bank can be judged from the amount of paper it uses in all its internal and external communications—900 million pages a year, or more than 90,000 per staff member.[61] Little can be kept secret for long in such a refuge of the chattering classes.

So to speak of a tacit agreement by aid workers not to raise fundamental questions, similar to the unspoken consensus that led to the Bre-X scandal, might seem improbable. But then again, there are a lot of well-paid jobs at stake.

Indeed, if you want to see the 'development community' in concerted action for once, just write a letter to an influential journal suggesting that the money that is being spent through them is doing no good. In December 1994 *The Economist* ran a story on a duo of papers from Peter Boone of the London School of Economics[62] that looked at aid flows to no fewer than ninety-six countries over the twenty years between 1971 and 1990. So it was not a matter of being selective about countries and periods, which the protagonists of aid often use to claim success. The results were frightening.

Infant mortality was taken as the best measure for determining whether conditions for the poorest people were getting better thanks to aid. The bottom line was that there was no statistically significant effect. There was, however, a relationship between the political environment and infant mortality; infant mortality was steadily lower in liberal democracies.

Dr Boone pointed out (defensively, in a letter to *The Economist* the following week, no doubt hearing rumours of the roasting awaiting him) that he had been misquoted, that health spending was only 6% of the total aid flow in the period surveyed and that his main conclusions were broader; that countries which receive aid at high levels (of the order of 15% of GDP) show no greater improvement in basic measures of human development or growth than countries receiving no aid, and that 'we should strive to make aid programmes really succeed.' This scurry for cover was understandable, for the next week's issue was potent with the denunciations of these conclusions from the great and the good of the development community, much as the Lutheran missionary Johann Rebmann had been maligned by experts who held it as self-evident that snow could not exist on the equator.

In fact, Dr Boone's use of infant mortality was an excellent criterion, as almost all the other sectors popular as a destination for development assistance had some bearing on it, whether the provision of rural clinics, of clean water, of improved crop yields, of women's education, of electrification, of telecommunications, of what-have-you. Whether he likes it or not, Dr Boone's analysis showed what is broadly agreed in private by all those who have

All poor together

been involved with development assistance in Africa; it doesn't work there.

But try and tell this to the people whose whole careers are built around the assumption that the development assistance business was doing good. Your head will bow, as did Dr Boone's, before a blizzard of indignant obfuscation.

Thus—'The facts reported in that article [an *Economist* account of an more upbeat UNICEF report] underline the recklessness of the assertions made in your earlier piece.' (James Michel, President of the Development Assistance Committee, OECD.)

And—'Almost any result can be obtained when aid is lumped together ...' (Jon Wilmshurst, Chief Economist of the British Government aid agency DFID.)

And—'You do not seem to recognize that the evidence on the impact of aid is now a matter of considerable research, much of which reaches different conclusions from yours ...' (John Howell and John Healey of the Overseas Development Institute, London.)

And—'Kingsley Martin, a former editor of the *New Statesman* once defined orthodox economists as "people who thought they could prove that it was socially harmful to give a penny to a possibly undeserving beggar." When I read your article on foreign aid, I thought "Come back, Kingsley, all is forgiven".' (John Toye, Director of the Institute of Development Studies, Brighton.)

So be warned; the future of whole institutions stuffed with experts is at stake when the usefulness of aid is questioned. And in the meantime, enjoy the spectacle of this author putting his head on the block in turn.

●   ●   ●   ●

At the end of the twentieth century the development assistance business for sub-Saharan Africa was spending in the region of seven billion dollars annually in grants and five billion in super-soft loans and providing jobs there for about a hundred thousand expatriates. Almost all of this goes in support of governments or their commercial arms. The latter, known as public enterprises or 'parastatals' (jocosely 'paralysed statals'), are usually national undertakings like airlines, railways and marketing boards. In some countries, for instance, Zambia in the late 1980s, these included ventures as diverse as copper mines, safari camps and hairdressers.

The ineptitude of states, particularly those in Africa, in making use of this wealth is no longer a matter for private and unofficial concern. State-run enterprises in particular have shown an extraordinary level of financial incompetence.[63] 'The actual effect of spending in almost every area of endeavour—from infrastructure to education—has been much less than the potential.'[64] As this quote from the World Bank shows, some sections at least of the 'donor community' are starting to wonder why aid does not work like it should.

After all, this effort has been going on for a quarter of a century or more, and during this period the standard of living for the average individual African has regressed to about the level he or she enjoyed in 1970 (according to the World Bank) or 1960 (according to OXFAM). By now much of the continent is in a state, as Lytton Strachey said of Keate's Eton, of anarchy tempered with despotism.

Reassuringly, a whole sub-discipline known as development economics has sprung up to deal with such anomalies. Its experiences, if not its advice, will be drawn upon shortly; in the meantime a look at the larger picture is useful.

The total foreign aid bill for the last fifty years is around a trillion U.S. dollars,[65] with Africa taking under 20% to start with and over 40% by now. Between 1981 and 1993 the per capita GDP grew at annual rates of 0.3% in Africa, 4.3% in Southeast Asia and 7.4% in East Asia.[66] Obviously then, not all of this money has been wasted. South Koreans started this half-century level-pegging with the people of Ghana in terms of income. At the end of it the Koreans were eight times richer than Ghanaians in real (purchasing power) terms, much more in nominal terms. Yet Ghana is one of Africa's success stories, having climbed back from a negative growth rate (of about minus one percent a year) in 1975–1985 to a positive one of about five percent a year in 1985–1995. Success story maybe, but Korea's annual growth rate between 1980 and 1995 averaged 7.5%.

In any event, to emulate Korea in Africa it is not enough to achieve Korean rates of growth. Korea's population stabilized in the 1960s while Ghana's in the 1980s continued to grow at 3% annually, making its 5% apparent growth an actual 2%. Korea in the 1960s and 1970s had substantial 'resource transfers' (investment plus some aid) amounting to about ten percent of GDP; Ghana's aid dependency (not at all the same as investment, as you must now realize) rose in the late 1980s to 8% of GDP; apart from gold mining there was little new foreign investment.

A 2% effective growth rate condemns Ghana to a fifty-year climb to achieve (today's) Korean levels of wealth. So in the mid-1990s the country set itself a growth target of 8% in order to achieve a 5% real growth rate, or one that is starting to approach Korea's. Ghana's ambition is to achieve middle income status in 2007, the 50th anniversary of independence. It will have to hurry; its GDP growth in 1996/1997 was only 0.5%.

The temptation to scoff is strong. Ghana manifestly will not become a middle income country by 2007; it will be lucky to achieve it by 2057. And whatever it is that causes a country to grow in Africa, it is not development assistance; the thirteen main recipients of aid in sub-Saharan Africa for the ten years between 1983 and 1993 (which included Ghana) had an average per capita growth rate of *minus* 0.2%. Aid's average contribution to their GDP was about 35% during this time. Dr Boone was right.

But scoffing will not solve the problem. Can development economics? What does it say? A wander through these outlying groves of academia may not be as dull as it might sound.

The first serious thinker on this subject was Joseph Schumpeter, with his book 'Theorie der Wirtschaftlichen Entwicklung' (Theory of Economic Development) in 1912, a book that focused on the role of innovators. In 1934 he proposed that development was propelled by the actions of entrepreneurs.[67] These naïve ideas, implying that development was largely independent of government intervention, were properly disregarded. As Keynes had noted, businessmen are driven by 'animal spirits', and it was therefore logical that they were not fit agents for the delicate processes of development. More importantly, apart from some non-governmental organizations such as charities, it was simply not politically possible to give the private sector the money; it had to go from developed country taxpayers via their governments to developing country governments for the latter's special use.

As a result, until about 1985 the branch was dominated by two contrasting strains, whose antecedents stretch back to an argument between Nehru and Gandhi in pre-independence India.[68] Nehru, in jail at the time, believed in rapid industrialization, if necessary through the nationalization of foreign-owned industries, as a way forward, so employing the burgeoning and impoverished peasant population. Gandhi thought that the use of improved technology would reduce rather than generate employment, and that the rural populations could be the losers. Both thought that whatever was done, the state should do it.

Nehru, of course, won. Ever since governments started shovelling their taxpayers' money into developing countries, the objective has been to replace traditional economies with those based on Western, or rather Northern, technology. (Yet in some cases, for example, the complex trading networks for gold and gemstones in Africa, it would be very wrong to speak of these systems as 'primitive'.)

Unfortunately, what was not considered was the possibility that these advances only worked when the 'Northern' environment, in which they had been developed and perfected, was in place as well. This environment includes workers whose morale is not destroyed by the demands of the extended family, opportunities for using the brains and hands of clever women as well as clever men, the rule of law, property rights, transparent administrations, honest management, good infrastructure, literate populations and so on.

However, at the time the convincing facade of the Soviet Union's command economy made these reservations seem irrelevant, if they were pondered at all. It was even contended (startlingly for anybody who has worked in business, even big business) that there was no reason why large state firms should be less efficient than private ones, since both were run by bureaucracies.[69] This aloofness from the realities of human nature, in particular from Lord Acton's dictum about the corruptions of power, is a worrying characteristic of economists; the possibility that state-run firms might be extremely vulnerable to political manipulation does not seem to have been seriously considered at the time.

So there it was; the 'Nehru model' of development economics was adopted, where rich countries transferred technology to poor countries, so that state-run mega-industries, secured by tariffs against foreign competition, could offer employment in the money economy to the ever-increasing numbers of subsistence farmers.

It was hopelessly wrong. It assumed that farmers in developing countries grew only just enough for their own consumption; but, to quote an informed observer, 'a true subsistence sector exists almost nowhere';[70] rural poverty is more the result of inadequate prices due to limited market access by the farmers, not because the farmers are content to grow just enough to live on. The theory also assumed that their numbers would continue to swell; all successful countries have brought their population growth rates under control. It further assumed that the right technology was of the big, smokestack variety; the hulks of these ventures, steelworks, glass-works, assembly plants, gold mines even, can be seen sprawled across the landscape in almost every developing country I have worked in. It assumed that the markets for their products were internal, yet every success story in development has been from success in international trade.

The failure of these ideas put development economics into denial. Obviously, it was said, it was not that they could not work, it was because the dice were loaded to stop them working. The developed world had control over many, if not most, prices. Trade barriers, along with quality and environmental standards prevented fair competition. Sometimes, certainly, there was blatant interference; Nehru harped on about the British in India in the past preventing Indian industrial projects from going ahead.

But the reasoning went further; these obstacles could be overcome and self-sustaining growth could be achieved only once the investment ratio reached a certain point. Capital accumulation was the thing; clearly more, not less, money was needed for take-off.[71] The situation resembled the tactics of the Allies on the Western front in the First World War; when the initial confrontation showed that a machine gun crew could stop a whole battalion, the only solution the generals could think of was to send over more battalions.

These arguments, plus the unanswerable one that development assistance is a necessary act of charity, sustained, even strengthened, the flow of aid money into government industrial projects, with remarkable results. As the *Wall Street Journal* noted in 1989:

> According to recent World Bank measures of industry's share of gross output, sub-Saharan Africa currently looks more 'industrialized' than Denmark ... As for the relative level of investment, World Bank estimates suggest this to be lower in West Germany than in such countries as Togo, Nepal, Egypt and Costa Rica. Recent estimates put the investment ratio in Rwanda well above that in Belgium; Mali's appears to be higher than France's and Sri Lanka's higher than Britain's. The investment ratio appears to be higher in the People's Democratic Republic of the Congo than in Japan and no Western country is estimated to have an investment ratio as high as Lesotho.[72]

The consequence of all this 'investment' in government projects, if not known to the reader before, will certainly be now. Whatever it did, if anything at all, it did not achieve 'take-off'.

Initially, the failure of these scuttles by the development community up and down the runway while flapping their arms were not seen as a consequence of government (or their) incompetence. Rather was it acknowledged that the objectives of growth through industrial investment, were wrong—which, fair enough, they were. This admission was spelt out by J. K. Galbraith, who said, apropos of his experience as America's Ambassador to India in the 1960s:

> ... we could supply capital and, in principle, useful technical knowledge. The causes of poverty were then

derived from these possibilities—poverty was seen as the result of a shortage of capital, an absence of technical skills.[73]

Until then, it had been a given that industrialization could remove poverty. Nehru, amongst other influential thinkers, thought that economic growth would cause all boats to rise with the tide. But what had been seen so far, in Africa at least, had been increasing differences in income distribution. The benefits of development assistance went largely to urban elites, while in the countryside there was increasing poverty because governments were controlling the price of crops to keep food cheap in the cities. Some of Gandhi's fears, it seemed, had been justified.

In the meantime, at least 25% of the aid money that had been poured into developing countries was in the form of loans; however 'soft' these might be, at some stage the money was supposed to be repaid. Again, almost all of this was from governments to governments, and again, Africa was the extreme case: by 1984 only 2% of its debt was to the private sector.[74] Further, despite the flood of well-intentioned development assistance into government efforts at industrialization, Africa had seen its portion of global manufacturing stay at around 0.4%. As part of the cost of holding this tiny, stagnant market share, between 1962 and 1990 Africa's external debt service had grown from a tiny 2% of annual export earnings to a ridiculous three times that income.[75] The numbers were by then academic anyway; the reality was that the continent had stopped servicing most of its debts.

When, at the beginning of 1980, amidst general indifference, the United Nations proclaimed its third development decade (the first had been kicked off by President Kennedy in 1960), a paper from the United Nations Food and Agricultural Organization (FAO) noted at the time that the rural world had more poverty, more under-employment, more malnutrition, more illiteracy and more poor health than ever before.[76]

So, back to Gandhi. It was concluded that the focus must be on poverty alleviation, not a race for growth.* Accordingly, a slew of poverty-reducing objectives were decreed under the general formula of 'xxx for all by the year 2000', where xxx was, for example, health (from the WHO), housing (from that branch of the UN known as Habitat) and education (from UNESCO).

It is redundant to say that all these objectives were missed; in retrospect they can be seen as acts of great silliness, doomed to failure and ridicule, a consequence of the enormous institutional and intellectual inertia of the development assistance business, its insulation from reality and the corruption and incompetence of the governments it nominally helps.

However, even while these fatuous targets were being announced in the early 1980s, a new consensus on what had to be done was emerging. To anyone outside the politically charged halls of the United Nations it was clearly no longer tenable to base development (or even development economics) on the forty-year-old ideas of a couple of Indian socialist statesmen. In certain countries growth was being achieved on an unprecedented scale without their advice.

For all the while that this debate was going on, the 'Dragons' of South East Asia were steadily powering themselves upwards. But apart from being export-orientated, with high investment (and high savings), frustratingly they seemed to offer no single model for their prosperity. Hong Kong and Taiwan, the most successful, were also the most open economically and grew mainly from the activities of relatively small businesses. Japan and South Korea were dominated by conglomerates, but Japan raised its finance internally (feeding off inflated property values) while Korea simply borrowed abroad. All that China, the biggest and poorest, did was to discard key portions of its communist theology to allow some property rights and entrepreneurial activity, whereupon it promptly achieved, and sustained, annual growth rates of around

---

* Despite this a hankering after industrialization was still evident; the United Nation's target growth rates for the Third Development Decade were 9% annually for industry, 4% for agriculture. No targets were set for the service sector, where real, if informal, growth was occurring in many developing countries.

ten percent. Economically it remained a closed (and tyrannical) nation. South Korea simply exported for all it was worth: its exports as a percentage of GDP went up from 3% to 30% between 1960 to 1990. For much of this time it remained a police state. Only Hong Kong, and to a lesser extent Singapore, Malaysia and Taiwan, were reasonably 'free' over this period. Most of the nations involved are densely populated and have limited natural resources, but Malaysia and Indonesia are the opposite. None of these states were unequivocal democracies.

Still, they worked, and from this came, in the early 1980s, a general agreement on policies that invoked the word *macroeconomics*. Because it was a joint World Bank-IMF strategy it became called 'the Washington Consensus'. In essence this consensus (the *paradigm* if you are so inclined; most development economists are) is that what has to be in place is fiscal discipline (i.e., low inflation), low taxes on a wide base, parastatals privatized, minimal government intervention in the market-place, no exchange controls, aid money not spent on political objectives and minimal regulatory obstacles. The implementation of these rules was known as *structural adjustment,* and its principal administrator was not the World Bank (for which it must now be profoundly grateful) but the International Monetary Fund. This was because there were no projects involved. 'Balance of payments support' is the term used for this generalized lending of yet more money to governments.

Anyway, the deal is this. Provided a bankrupt government agrees to abide by the macroeconomic rules of the Washington Consensus, the IMF will lend money to it, and if the IMF lends money the rest of the development community will weigh in with their soft loans and grants. It was proudly claimed that Jerry Rawlings of Ghana, Yoweri Museveni of Uganda and Laurent Kabila of the Democratic Republic of the Congo are amongst the converts to these ideals.[77] Good for them.

There is, however, no certainty that structural adjustment works. Certainly it will succeed in countries where there is what is wistfully known as 'a good institutional environment'. But they are very few, and usually they have no need of structural adjustment programmes anyway. So for all the rest, those with varying degrees of corrupt and inefficient administrations, a sort of mating dance occurs. The country needing the money undertakes to implement reforms, the IMF, etc., starts to disburse. Whereupon the zest for reforms disappears (it usually requires sacking most of the president's relatives, cronies and long-time political mates). The IMF goes into a huff and refuses to send the next tranche, and the other donors then do likewise. A stand-off ensues, during which the country suffers and the man-in-the-street gets poorer. Eventually the government agrees to implement a revised, softer, programme (for by now the absence of hard currency inflows have made the original inflation targets right out of reach, the government's budget deficit is way off target and renewed exchange controls have become necessary). So then the IMF disburses again ...

This dignified cavort has become almost institutionalized itself. Zambia, for instance, had no less than eighteen tranches of IMF balance of payments support (between 1966 and 1969 and between 1990 and 1993) during which time its macroeconomic indices—of trade, of fiscal and of monetary policies—got worse and worse.[78]

This might not matter overmuch — Zambia? Where's that? — what is much more important is that the Russian bear—along with those from the Ukraine, Belorussia and the rest—has also been slowly capering about with the IMF and Co. Here the numbers are really unsettling; Zambia has accumulated a debt to the IMF of something over a billion dollars over thirty years, but U.S.$50 billions of IMF money has vanished into Russia—quite literally; there has been no perceptible effect—in under a decade. Was the Washington consensus doomed to go the way of Nehru's ideas?

Certainly by the end of the century it was in tatters. First came the failure of its role models during the East Asian financial crisis in 1997. Then, after a similar set of barely-averted crises in South America, the World Bank's chief economist, Joseph Stiglitz, began publicly sounding off against the consensus, or more accurately, against the IMF's implementation of it.

He said that the focus on macroeconomics was wrong; that

without a proper and working regulatory framework it would fail. Freeing up such economies simply made it easier for the richer, cleverer inhabitants to establish private monopolies in place of government ones and then spirit their assets abroad, just like, well, Zambia, for instance. It was a hugely embarrassing moment and, true to form, there was an immediate, wrathful response. One distinguished economist was quoted as saying of Joseph Stiglitz that, 'Without knowing anything, he mouths any stupidity that comes into his head.'[79]

Certainly, the Stiglitz heresies were remarkable. The nurse sent the bath water hurling out of the window, possibly with a number of babies: in a speech in Helsinki, Finland, in January 1998, he committed heresy after heresy:

> ... on inflation below 40% a year, there is no evidence that [such] inflation is costly.
>
> Ironically, macroeconomic stability, as seen by the Washington Consensus, typically down-plays the most fundamental sense of stability: stabilizing output or unemployment.
>
> The advocates of privatization overestimated the benefits of privatization and underestimated the costs.
>
> The unspoken premise [of the Washington Consensus] is that governments are presumed to be worse than markets.
> ... I do not believe [that].

Anathema! anathema! Not unsurprisingly Stiglitz left the World Bank in late 1999 after only two years in the chief economist's job.

The IMF struck back in dignified fashion, simply referring to their research that showed that, given a 'good institutional environment', structural adjustment worked. These rebuttals were not reported as thoroughly as Stiglitz's speeches, perhaps because of the opacity of the language in the IMF papers:

> Quantitative measures of policy complementarity are developed, and the study shows empirically, through both

an outcomes-based probability framework and a standard regression analysis, that these complementarities are significant and robust in explaining growth outcomes over the period 1985–95.[80]

And that is only the abstract.

In a remarkable speech back in 1987 the resident representative of the World Bank in Zambia, a Nigerian, spoke with unprecedented candour. It is enough only to give one paragraph:

> The IMF and/or World Bank reforms which many African countries are now undertaking are only some of the things that need to be done. They are designed to help us stop shooting ourselves in the foot economically, and to start undoing some of the economically harmful policies of the last twenty years. They are things we Africans should be doing for ourselves without the help of the IMF or the Bank. We are shareholders (i.e., part-owners) of the both the IMF and the World Bank. They cannot enter our countries without our consent and indeed, very often at our invitation. Bad-mouthing them is therefore rather childish.[81]

There were no reported responses to this advice, and by now most African countries no longer have the chance of taking it; their debts have reached a point where, invited or not, the IMF and the Bank are the only girls visible in the ballroom. In addition the crowd around the exits is in a threatening mood. The best hope for poor countries seems to be to exploit the rift that is opening in the consensus to get in some generalized lending from other donors without IMF approval.

African leaders have been quick to cotton on to what has happened, and to exploit it, chiefly by praising the Bank at the expense of the IMF. At a time when Zimbabwe's inflation rate had increased from 25% to 70% in the space of a year, and the Zimbabwe army was bogged down in the Democratic Republic of the Congo whence he had sent them, President Mugabe of Zimbabwe said:

All poor together

The IMF says you cannot borrow beyond a certain target but the government cannot be expected to make the money by mysterious means ... [if the government borrows for productive purposes] I do not see any reason why there should be limits in the government debt. Government must continually borrow ... There must be more government bonds but these offend the IMF since the IMF is a god we cannot offend [sic]. I think the god has to be offended because it is a false god.[82]

Robert Mugabe was perhaps buoyed by the fact that he had just got back from a visit of mutual self-congratulation to one of his few remaining fans, Fidel Castro of Cuba. But he was treading a naïve and dangerous path, because by further stoking the fires of dissension he was damaging the credibility of the whole aid process, not just weighing in on the side of the World Bank.

For this fundamental argument about structural adjustment has enabled the political opponents of aid (who are no longer principally American, and no longer exclusively right-wing) to argue that as the specialist agencies themselves don't seem to know how to make aid to Africa work, it would be foolish to continue handing it out. After all, up to now, judging from their fundamental differences, everything spent so far may have been wasted ...

The debate compounds an earlier one, over what to do about Africa's debts. These, largely accumulated over those three 'decades of development', have become intolerable, yet short of letting developing countries walk away from them (politically very difficult), there seemed no simple solution. The wrangle, again essentially one carried out in developed countries between and within governments and the development agencies, had already made such governments unwilling to commit more project aid, as that would simply increase such debt. In the meantime, as *The Economist* observed, 'the magic of compound interest has cast such an evil spell that, in sub-Saharan Africa, for example, two thirds of new sovereign debt since 1988 has been incurred to refinance arrears and capitalized interest.' [83]

In September 1996 James Wolfensohn, the President of the World Bank, breathlessly announced what seemed to be 'a breakthrough ... It is very good news for the poor of the world.' This was the HIPC debt initiative, where the abbreviations stand for Heavily Indebted Poor Countries. These were the no-hopers; countries that might never be able to get on top of their obligations. There were four in South America (Bolivia, Guyana, Nicaragua and Honduras), three in Asia (Myanmar, Laos and Vietnam) and no less than thirty-four in Africa.

What is ironic in the light of the row over whether structural adjustment works is that the HIPC mechanism requires an existing programme of 'strong policy reform' (that is, vigorous structural adjustment) for aspirant debtors to qualify in the first place, and then three years thereafter of the same medicine. If at the end of that period, the countries involved had managed to reduce their debt to a level where it would be sustainable after a further three years of such rigour, well, that was OK then; they would be left to get on with it. If they hadn't, then an analysis would be made of what degree of debt relief would be required to achieve sustainability in three years. Then, if the creditors agreed, the required amount of debt would be written off, provided the offender then observed a second three-year period of 'strong policy reform'.

We have seen that even the IMF acknowledges that for reform to work there must be a 'good institutional environment', and that this, despite promises, promises, is a rare thing in Africa. As a result by 1999 the HIPC debt initiative had become a wallflower in the IMF's ballroom. Only two countries had actually benefited (Uganda and Bolivia) and seven countries had been assessed as qualifying. But as the creditors were still arguing about the fine print, it was looking increasingly unlikely that the scheme would have much effect on the U.S.$170 billion that the countries collectively owed.

Thus the HIPC initiative. But to return to the much-abused Peter Boone and his discovery that there was no correlation between aid and infant mortality. In the middle of 1998 two World Bank researchers came to the conclusion that it was true—in countries with 'weak economic management'.[84] As this frailty applies to most of the third world, not Africa alone, it is little

All poor together

wonder that there is no link.

'Weak economic management' is a broad, wishy-washy and perhaps misleading description anyway. The one conclusion that development economics should have made from its experience of fifty years is that it does not really matter whether you have a closed economy or an open one, a fixed exchange rate or a floating one, whether you control the economy by mainly monetary means (setting interest rates) or mainly fiscal means (setting tax and tariff rates). It may not even matter that the dead hand of government is involved. These are all subsidiary to the essential requirement that your people are able to work and innovate in an environment that makes them capable of competing in the international marketplace.

So, again, this is the prevention of the impoverishment of workers from the demands of large families, the freeing of women from continuous child-bearing so their brains and dexterity can be used, the presence of the rule of law, of secure private property, of a limit to the degree of corruption in the regulators, of employers who themselves are not crooks, of transport, electricity and water supplies that work and a reasonable level of literacy. Not too much financial and monetary policy—macroeconomics—in all that.

Of course, this wish-list amounts to a level of freedom from cultural and administrative constraints that most poor peoples do not have, certainly not in Africa. But try reversing the argument. Instead of arguing that it is necessary that institutions have to be good for reform to work, why not accept the reality that they are not, and will not be for the foreseeable future. Then ask 'what outside intervention is needed to allow the majority of people in a country to become wealthier even if the institutions are not good?' Immodestly, this book hopes to show how to achieve that.

Our detour through the recondite pathways of development economics is over. Readers will be pleased to hear, after all this carping, that an authoritative review of the field has said that 'Development economics has made remarkable progress in the past fifty years.'[85] That is not, as will now be understood, the same as saying that development itself has progressed correspondingly, but it is, perhaps, a step in the right direction.

Chapter Thirteen

SMALL, SMALLER, ARTISANAL

Small is Beautiful
— Title of a book by E. F. Schumacher, 1973.
... but big pays more
— D. James, *A Spy at Evening, 1991.*

The time has come to step down from the critic's seat into the arena, and to tell of a practical example of how trying to do good in the aid business can end up crippling the economic development of the very people it is supposed to serve.

• • • •

Ankara was hot that September, an unlovely city, like Cairo with its serried concrete towers of worker's flats. Only around the diplomatic quarter, the government offices and the university were there anything like enough trees. Every now and then the sweltering, seething city shook a little as the Anatolian plate grated against the Eurasian one. Improbably enough, at the time—1988—plate tectonic theory was so fresh that almost nobody outside a few geology departments realized this. (But then again, it seems improbable now that it was not until well into the twentieth century that astronomers realized that ours was not the only galaxy in the universe.)

Plate tectonics is said to have a Zimbabwean, or more correctly, a Rhodesian connection. In the 1930s a young American geologist called H. H. Hess began his career working for a mining company there. The story goes that he was set to using the technique of loaming, which involves panning soil samples across a line of country, in the hope of picking up the few colours that could indicate a hidden reef underneath. It was the precursor of easier, more sensitive techniques using cyanidation and solvent extraction.

All poor together

Loaming is a back-aching activity that becomes pretty mindless after the first few days, but Hess managed to stick it out for a year before giving up in disgust.

During the Second World War he was in the U.S. Navy, and noticed that the sonar echoes from the sea floor (he was supposed to be hunting for submarines) suggested a massive system of parallel formations of north-south volcanic ridges coming from somewhere in the centre of the Pacific Ocean. He suggested that these ridges were being created in a somewhat spasmodic way—hence the parallel formation effect—and that when this moving ocean floor meets the continents it dives down again to the earth's mantle beneath them.

It was a theory that brought together some ideas that had been around for some time, from, mainly, southern hemisphere geologists. The confirmation came in the 1960s when it was discovered that the earth's magnetic field reverses every million years or so, and that this reversal is recorded in the ocean floors as a series of stripes parallel to the central ridges.

Hess stayed with the Navy. It is said that when a research student crashed a jeep, Hess ordered him to push it into the sea and report it as 'sunk'. This was something, he pointed out, that the bureaucracy was well adapted to deal with, whereas any other description of the loss would lead to endless official nagging. It was no wonder that he retired an admiral.

I was here, in this wobbly city of Ankara, under the aegis of one of the world's great bureaucracies. I was at a United Nations conference, a relatively little one, and so appropriate to its subject, which was to do with small- and medium-scale mines, or mining, or miners, I forget which. The 'medium-scale' comes into it because small- and medium-scale enterprises (SMEs) have been a target of development assistance affections for many years, to the point where the abbreviation carries with it useful purse-loosening connotations that merely 'small-scale' does not.

These little tiddler conferences have a pattern, which is like this:

There are national delegates from perhaps thirty developing countries who get up in turn and tell you what their countries are

Small, smaller, artisanal

doing about (in this case) small- and medium-scale whatever. Then there are the United Nations people from the appropriate department who get up and go on, and on, about their organization's invaluable services and achievements. In this case the department no longer exists (it was something to do with mineral resources), extinguished by a peculiarly successful piece of re-engineering developed by the UN. This technique was to organize unwanted appendages under a weight of abbreviations so long and incomprehensible that they deter any communication, and the limb withers away. The last I heard of that lot was that they had been consigned to the STEENRD of the DESD.

After these head office people rank the interpreters, whose self-importance blossoms on these occasions. They are like me, persons dependent on the occasional windfall rather than a steady income. However, moderation seems to be rare in polyglots, and once ensconced in their little boxes they are avid clock-watchers, marching out in a marked manner as soon as their time is up, so that the conference collapses in confusion, with baffled and indignant delegates taking their earphones off and staring angrily around, looking for a scapegoat. Then frantic negotiations have to be conducted in semi-public to get the interpreters back up again. Perhaps if I was doomed to deal with the verbiage from whole halls full of gasbags, I would also be bloody-minded.

Finally there will be the resource persons. *Consultants* was evidently deemed too constricting a term or perhaps it didn't translate well in Pushtu. Anyway, in Ankara I and a couple of others were resource persons, being quite well paid for the lectures we were to give.

What had not been made clear until I got there was that I was supposed to have brought a paper for distribution as, it turned out, everybody else had. It was not a mistake I made again, for I spent most of my five days in the city before a computer, frantically typing it up. I missed a lot of the conference, although I now know that this was not a major setback.

However, I completed my work in time to attend the final two days, so that is enough for the intro as to why I was there. Let me get to the point.

All poor together

When I got back the entire conference was devoted to defining what was the maximum size of a 'small-scale mine'. Surprisingly this was a matter—sorry, an *issue*—which aroused such emotion that eventually the whole of the final two days were given over to heated arguments about it. These discussions were to me, a mining consultant from a mining country, incomprehensible. There was no distinction made by size in our mining law, or that of successful mining countries like Australia, Canada or America. Everybody worked under the same system of pegging claims, whether peasant or potentate. It was not practicable, anyway, to split up the industry into large and small. A mine that hoisted a thousand tonnes a day would be small by coal mine standards, but very big if it was a gold mine.

The impracticalities did not end there; as soon as one starts on this path a major semantic slough halts all progress. There were medium-scale and small-scale mines, which seemed to be simple enough, even if not definable. But then, what were small mines *per se*? And what were—a phrase I had first heard Dr Kamel use a few years before—*artisanal* mines? It was no good just saying that you knew one when you saw one; there were, it seemed, functionaries all over the third world with rubber stamps poised, waiting anxiously for the criteria we were arguing over.

The controversy became heated, the hours dragged on, the interpreters flounced out. In the end, out of exhaustion, the conference voted to adopt a number already fished out of the air at a previous meeting in Lima, Peru. This was that a small mine (never mind the -scale suffix or the vexed meaning of artisanal) was one that hoisted less than 50,000 tonnes a year. Its sole justification was that it had been used for many years by the *Mining Journal* group of magazines to distinguish the noteworthy mines from the rest of the field. As such it had no relevance at all to the widespread informal pitting and panning that was going on all over the world.

Thus as an exercise in futility those two days were unsurpassed. Yet it turned out that this was an ongoing debate, left unfinished from the previous gathering. Indeed a similar fatuous discussion had surfaced at another, unrelated, small mines conference I had

Small, smaller, artisanal

attended in London a year or so before. At the time I had dismissed it as an aberration because Britain had ceased to be a serious mining country (it has lots of quarries, as the Brits keep pointing out, but they are really not quite the same thing) and has planning laws directed at discouraging such activity, as well as an archaic and unsatisfactory mining law.

My own contribution in Ankara was limited to suggesting that any mine without a defined ore reserve was a small mine and asking why was this nonsense (I didn't say that) necessary? My *interventions*, as UN-speak has it, were met with some bemusement. It was politely, carefully pointed out that, of course, definitions were necessary. Governments needed to know.

I gave up, mystified. Why did they need to know? We were evidently on different wavelengths.

The mystery was slow to clear, although over the years I managed to become something of an authority on small mines in Africa in the eyes of the United Nations and the World Bank. This is a mistake; I am actually an authority on why there aren't any to speak of (except in freak cases like Zimbabwe).

Anyway, it emerged that during the 1960s and 1970s most African countries had their mining laws rewritten. Apart from the usual overhang of aid looking for someplace to go, and the common desire to tinker with anything old, working or not, there were three factors that contributed to this legislative fashion.

First of all there was the demand from the new governments that they reassert sovereignty over their national resources, secure their economic as well as their political independence, or some similar slogan that got out the knee-jerk patriots amongst the voters. This coincided with the leadership of a very large number of these newly independent states becoming attracted to a socialist or communist—in any event authoritarian—system of government.

Because the adoption of these policies cut them off from most sources of international risk capital, and because such governments saw it as proper to control all economic activity, it was necessary to construct siege economies with price controls, and a ring fence of currency controls to prevent capital flight. Inevitably this resulted in the creation of vigorous black markets in hard currency.

All poor together

Then the gold price was freed up in 1971 and went scuttling upwards. The parallel economies that developed meant that those lucky enough to have access to foreign exchange could afford the luxuries (and often necessities) that were unavailable to the rest of the population. Gold and gemstones are, in effect, hard currency hidden in the ground, spurring 'barefoot prospectors' to discover them.

This they did, in astonishing amounts. A combination of a high gold price and bad mining laws made artisanal or small-scale gold mining one of Africa's few growth industries. A very rough estimate of the yearly tonnage of gold they now produce can be made from a review of the global supply and demand situation put out every year by Goldfields Mineral Services in London. Called 'Gold 1996' or whatever the year is, this is the principal source of information for where gold is coming from and where it is going to, and for how much.

Because—amazingly—there are only something like a score of significant, formal, operating gold mines in Africa outside Zimbabwe and South Africa, the difference between their output and the total gold coming out of Africa is the artisanal/small-scale output. This is a rough figure, and something of an under-estimate, as much goes to India without the formality of being recorded. In 1986 this calculation (an average of the past three years actually; the figures were very jumpy) gave 20 tonnes; by 1996 it had risen to 84 tonnes. At the prices prevailing in the latter year this gold would have been worth over eight-hundred million dollars.

For diamonds the quantities are less certain, but one set of figures[86] suggests that it is of the order of three million carats a year. Of this about two million carats comes from Angola alone, and the quality of the stones means that their value is several hundred million dollars annually. All in all, well over a billion dollars of gold and gemstones leave Africa illegally every year.

Given this figure, the mining population can also be roughly estimated. The distribution of income across a typical group of workers on alluvial and eluvial gold will be very skewed, as a classic paper showed long ago for Tanzania's Lupa goldfields.[87] The data obtained and the conclusions arising from this reference

Small, smaller, artisanal

are both curious and important and there will be more about it later.

The point is, despite many tales of sudden wealth, average earnings will be quite modest. The World Bank's World Development Report for 1997 says that almost half of Africa's people live on less than a dollar a day. This correlates with the largely current evidence about artisanal mining incomes on the continent. While these vary from country to country and group to group, such miners in Africa—mainly young, unemployed men and the wives of poorer peasants—appear to be prepared to persevere in the rigours of this sector if they anticipate getting an income equivalent to more than about U.S.$35 a month. In very crude terms this suggests that as many as four million people are engaged in artisanal mining in sub-Saharan Africa. Indeed, allowing for the seasonal nature of much of this activity, the population actually involved is probably much higher.

These numbers imply phenomenal growth, given that at most countries independence in the 1960s such artisanal mining as occurred was of a scattered and traditional nature (for instance, the women river panners *au pays Lobi* in the south-west of Burkina Faso, the diamond hunters of Sierra Leone and the 'galamsey' of Ghana).

Government reaction to this sensational expansion was usually unequivocal; the state owned all mineral rights, so artisanal miners were stealing from the state. The state had control of all foreign currency, so artisanal miners were, well, undermining the economy. In the convenient shorthand for all this, acquired from the Eastern Bloc, they were *saboteurs and racketeers*. Any step short of driving the miners from their diggings and arresting as many as possible would be weak-willed temporizing. Direct force was therefore employed in countries as diverse as Zambia, Tanzania, Mali and Ethiopia.

Of course, once the security forces had left the miners would invariably return, and output soon recovered to previous levels. Such were the rewards of artisanal mining that it was usually found necessary to station permanently some sort of armed force—police or para-military or army—in the mining areas. The common result was, as I had found at Buckreef, that the security forces themselves became involved in artisanal mining, usually striking deals with the miners on a tribute basis.

Faced with this embarrassment numerous governments asked for assistance from the donor community. It was hardly an area in which most donors had direct experience; indeed many of those who became involved had only minuscule mining industries of their own. However, as it was depicted to the donors as a legal problem, and as they had ample resources of lawyers, a large number of legislative studies were made, often in conjunction with an overhaul of the mining law inherited from colonial times.

The general conclusion of all this work in perhaps a score of countries was that artisanal mining and large-scale mining were incompatible. The received wisdom became that artisanal mining was 'a process set apart from the twentieth-century concept of mining'.[88] As such it had to be guided by government. The reference given above further describes the then new Tanzanian Mining Act of 1979 as a 'good example of a modern mining law which tackled the problem of small-scale mining'.

In this law the Minister of Mines (of Minerals, of Natural Resources, or whatever) is empowered to designate areas for small-scale mining only. However, if a small-scale miner discovers a large deposit then the mechanism provided by the Act allows his or her claims to be cancelled by the minister, by revoking the designated area.

The self-congratulation amongst aid agencies that ensued from this type of law probably arose from the enthusiasm with which it was received. However, this enthusiasm may not have been so much for a fine bit of legal drafting as for the immense discretionary powers it presented to the minister (or his nominees). The more opportunities for decision-making, the greater the scope for exacting a reward from the supplicants.

These laws also—and this is more to the point—fatally undermined the value of a mineral discovery for local people. It also assumed that the (generally poorly paid and unskilled) government geologists were capable of selecting, in the words of the Tanzanian act, 'suitable areas for small-scale miners'.

Small, smaller, artisanal

So what then is a 'suitable area'? Well, clearly, somewhere where there is little chance of a major mine being found. All right, here are three good possibilities. Firstly the Turkana region of Kenya, where the last Bulletin of the Kenya Geological Survey for this area declared that no gold showings had been found. Next, the Niassa Province of Mozambique, on the border with Tanzania, where the Portuguese geological maps, such as they were, showed the mountains there as being of basalt, a volcanic rock that never normally carries gold or gemstones. Finally, the Songea region of southern Tanzania, largely sedimentary rocks in which no significant mineral discoveries have ever been made (apart from a bit of coal).

Clearly, if one packed the artisanal miners off to these out-of-the-way places, they wouldn't get in the way of big, foreign, *legitimate* mining companies. Would they?

Yes, they jolly well would. About 1993 the Turkana people, reduced by banditry and misplaced aidery to near-destitution, began in desperation to scratch for minerals in their soil, and found, would you believe it, gold. Since that time the region has been producing over four kilogrammes of gold a week—a quarter of a tonne a year—and the Anglo American Corporation has taken an exploration concession over it. It appears (see the chapter on the Bre-X scandal) that there might be an epithermal deposit there; a first for Africa.

Then in Mozambique's Niassa province there was a major discovery of gold in those 'barren basalts' in 1992, by Tanzanians drifting south across the border, from where they had been successfully washing gold from the rivers. They were following the Livingstone Mountains, which edge Lake Malawi at this point. There is gold there, and there is also gold in the same mountains across the invisible frontier acknowledged in an Anglo-German agreement in 1886, the one that set out 'spheres of influence' in East Africa.

It is incomprehensible how the rocks were so badly misidentified, for these mountains are not fine-grained basalt but coarse-grained granite, emerging through a greenstone belt, the classical environment for gold discoveries. Nor does the geology

All poor together

change at the border; the Portuguese geologists had simply got it very wrong. Now several foreign companies are working in the area, which has been comprehensively divided up into concessions.

Then geologically boring Songea in Tanzania. In 1994 there was a discovery of alluvial gemstones by local people—sapphires, principally—of a quality and quantity sufficient to threaten to unhinge the world market. These had been washed down over the aeons from a source somewhere near the Selous Game Reserve, far to the north.

So it is going to be very hard to pack off small-scale miners to 'suitable areas'. Put them somewhere ostensibly barren and they will still find something to attract the foreign miners.

The upshot of these well-intentioned changes to third-world mining laws was to segregate the local miners from the foreign investors. Not very efficiently, for as we have seen the definitions were being hotly argued over many years later, but very unfairly. Local people who made discoveries were not allowed title to what they found, and were forced to hand them over to foreigners who were seen as better equipped to develop them. The idea of all mineral discoveries being the tradable, transferable property of the original finder was nowhere—unless that finder was a foreign mining company.

In practice the mining law was usually irrelevant; rural people in failing economies did as the Turkana and started to look for something to mine profitably, regardless of geological probabilities or what the rules said. It was that or starve. Hence the explosion of artisanal mining in Africa since 1970 and the consequent stream of new discoveries.

I had little idea when I arrived back there in 1982 of how well Tanzania could serve as an example of this egregious outlook. Truth to tell I wasn't very interested. There was something wrong, otherwise there would be a lot of little formal gold mines on the scores of good reefs the artisanals had found, but so much was wrong in the wretched country that it was hardly worth concerning oneself with the fate of one small sector.

Yet the new, foolish mining laws of the Tanzanias of Africa were surpassed by some of the francophone states there. Not content with the rewriting of already restrictive mining codes, they sought to separate out not just small-scale mines but medium-scale ones as well.

To achieve distinctions between these three different types of mines it was necessary to be very specific about the equipment used, or, more correctly, not to be used. In Mali, nine equipment items are specified for the *orpailleurs* (artisanal miners) to use on their *orpaillages*. Not all these are hand-operated; they include electric generators, pumps, mechanical hoists, compressors and jackhammers. However, the list does not mention wheelbarrows, tracked wagons such as cocopans, ventilation fans, drill-sharpening machines and so on. Mills (*broyeurs*) are OK. but whether this encompasses all the crushing, grinding and extraction activities that go on at surface is left unclear.[89]

Petite Mines are one up from *orpaillages*, and in the case of gold the reserves must be under or equal to five tonnes (but the reader will know by now that the reserve of any mine, let alone a *petite* one, is always a slippery phantom) and with a capacity under or equal to half a tonne a year. That is a lot of gold; out of Zimbabwe 1,500 or so producers only about 20 exceed this figure.[90] The net effect was that Mali, along with most of the rest of Africa, had only two types of mines, hundreds of *orpaillages* (for most of which the equipment they were permitted to use was an impossible dream) and a couple of massive foreign-owned enterprises.

The failure of this sort of unworkable special mining law for small mines occurred at a time when indelible scenes were coming from Serra Pelada in Brazil, of human walls of mud-coated miners struggling up the sides of the pit with their sacks of gold ore. In addition, concern over the environmental impact of small-scale mining was mounting, in particular about the long-term effects of mercury. Pollution was reported from a world-wide range of countries[91] leading to further calls to resolve the situation.

Consequently, there was a steady pressure on the United Nations and other relevant bodies, not for a better code—they already had that, didn't they—but for a workable definition of what small mines were. Hence the protracted, heated and ultimately

All poor together

futile discussions in places like Lima and Ankara.

Up to that stage I had never paid much heed to our mining law in Zimbabwe. It was a given. If you wanted to take a gamble in the mining casino you bought a prospecting licence for a few dollars, fossicked about for a bit (having first told the landowner, remember?), pegged the reef or whatever that you had found, and then set about exploring, developing or selling the property. As we have seen, this had given little Zimbabwe the largest mining sector in Africa outside South Africa, whose geology is freakish anyway.

Rhodesia's protracted rebellion (it lasted from 1965 to 1980) gave it one advantage. By the time we emerged from it the impetus to rewrite the mining laws of developing countries had been lost. It would be nice to think that this was because the new codes had manifestly not achieved anything, but a knowledge of the way aid works suggests that it was more likely because the fad had passed.

In any event we escaped the damage to the mining sector wrought elsewhere by a local political elite working in conjunction with an imported legal elite. The type of mining code they came up with works like this:

- you can only prospect legally in an area if you are a foreign company, and to get that you negotiate with the minister or his nominee;
- you can only get the right to mine a discovery if you are a foreign company, and to get that you negotiate with the minister or his nominee;
- you can only sell your discovery if you are a foreign company, and to get that you negotiate with the minister or his nominee.

However, subject to ministerial approval (or, of course, that of his nominee), you can, as a local inhabitant—a national—have access to these rights. In practice this means that you are either a close relative of the minister (alternatively of his boss, the president), or an agent working on his behalf. The Turkana discovery fell very quickly into the hands of President Moi's man of business, the minister of mines in Mozambique was the effective owner of much of the Niassa discovery and half the Tanzanian cabinet seemed to have a stake in the Selous sapphires.

Of course, as we have seen, there will normally be a part of the code that allows local miners to scratch away on sufferance, but they will be thrown out if they find something really valuable.

So, with a couple of exceptions, local people in African countries have no legal right to any mineral discoveries they make for themselves. As it is not theirs to sell, they either have to obey the law and walk away from it, or damn the consequences and try to exploit it with their own negligible resources: hand tools at best, often locally made. They become artisanal miners, trapped into the most primitive form of mining by the laws of their country. The definition of this activity, then, is what the United Nations' conferences sought.

In some countries (Ghana, Mozambique) they can own their discovery, but they cannot sell it. Still the trap, but a legal one. Because they do have the right to mine, such operations can become quite big, and the owner (who might be the finder, but who is usually somebody with the considerable resources and patience, not to say influence and muscle, needed to lodge a claim in Accra or wherever) will sub-contract sections of the mine to others. However, there can be no thought of putting up a proper mine. These, then are small-scale miners.

So that's it. Artisanal miners illegal, small-scale miners legal. What they have in common is a mining code that shuts them out of the market place in mineral titles that is to be found in all mining countries worthy of the name.

This is much more important than it might seem. It is simply not possible to mine properly on the cheap. To understand this, see what happens to the small-scale miners, who have no choice but to try to do so. In the case of say a gold reef, self-contained gangs dig pits of about a metre in diameter side by side along the strike (a different sub-contracting gang on each pit) down through the soft weathered upper rocks. The reef is brought to the surface and hand-sorted for visible gold; as a rule-of-thumb this will have about an ounce a tonne in it. The rest is often discarded for later treatment, while the good stuff is crushed in primitive pestles and mortars and the product panned by women.

The miners typically dig down through fifteen to fifty metres of the weathered near-surface rock until they come to the secondary enrichment. Here they burrow around a bit, often breaking through into the next-door neighbour's pit. It is at this point that the entombment and deaths occur from what in mining parlance are 'falls of ground'. Accidents in pits that are still being sunk are rare because reefs are usually close to the vertical and, unless the ground is very loose, such pit walls tend to be self-supporting.

The point is, once the rich level has been cleaned out, the miners have to start all over again, back on the surface digging another pit. This is because below the secondary enrichment this mining technique is uneconomic, certainly in comparison with the relatively easily-won gold above.

For one, the secondary enrichment is usually at the lower limit of the soft weathered rock. Without explosives the labour involved is five or six times that used above this level. Also the water table is often found at this point. Quite often there will be an entrepreneur—frequently the boss of the diggings—who can hire out a pump, but this is usually a cheap little diesel or petrol irrigation unit, the only sort locally available, not the high-lift submersible electric pump (needing a generator) required for safe and effective underground use. Finally, from here on down the grades usually never approach those in the secondary enrichment.

Suppose, notwithstanding the above, a gang of artisanal miners decided to mine below the weathered zone. To do this would therefore require a jackhammer to drill holes for the explosives (and compressor, compressed air line and clean water supply), explosives to go in those holes (and somebody who knows how to get them, store them and use them), a ventilation system to bring air to the miners (and for clearing the poisonous fumes from blasting) and usually the right sort of pump (with a power supply and a pipe to the surface). At this depth it is also desirable to have a decent windlass, with cable and bucket, to haul out the ore.

All this stuff simply will not go into the bottom of a hole a metre or so across which is already a tight fit for the miners themselves. Most important of all, pitting—or shaft sinking as it will become—is a very expensive way to mine. Miners—proper ones that is, neither artisanal nor small-scale—never work this way. They *stope* the ore.

Small, smaller, artisanal

In its simplest form stoping involves miners first digging one biggish shaft to about a hundred metres depth and putting a pump in the bottom. They then drive out from the shaft along the strike at two levels (say thirty metres and sixty metres) for about a hundred metres and connect them with a raise at the ends. Then they start blasting out the raise down into the lower drive, working backwards towards the shaft. After each blast, when the air has been cleared using powerful fans, they go into the lower drive and tram—a mining term for transporting rock; wheelbarrows can be used to do the tramming—the broken ore to the shaft for hoisting.

This blasting downward business, using gravity to break out large amounts of ore at one time, is stoping, and the big hole it leaves along the strike is the stope. The preliminary work that went on before to set up the stope is known as development. Development ore costs between two and four times as much as stoped ore to mine. Artisanal and small-scale miners are pit-diggers; they only mine development ore. So they are in a financial trap as well as a technical one.

There are other technical snares awaiting them. Their pestle and mortar crushing system will not grind up the ore fine enough to free all of the gold, and a lot—typically about 40%—is left in the tailings. There are fortunes to be made cyaniding the artisanal mine dumps of Africa.

To grind up ore fine enough to recover most of the gold requires a grinding mill. The smallest such will treat about ten tonnes a day, and that very inefficiently. Twenty tonnes a day is normally the economic minimum. But a metre pit ten metres deep with four workers will, at best, produce about two tonnes of ore a day.

About the only possibility here is to use what is know as a toll mill, where the miners send their ore. Indeed there are a number in Zimbabwe and one or two elsewhere in Africa. However, there is no practical way that a small pile of ore can be valued and paid for prior to getting the gold from it. The miners are certain that their output is full of nuggets, the mill owner is sure (and with more reason) that it is not. Even if assay facilities were available, accurately sampling a little heap of gold-bearing rocks, each a few

All poor together

centimetres in diameter, is not usually feasible unless there are absolutely no nuggets at all, not even the smallest. So the owner of the mill has to hire the whole thing out on a custom basis to the miners—these are custom mills not toll mills, strictly speaking—entailing frequent stops and starts. After each batch the miners spend much time cleaning everything out to make sure that they have not left anything of value behind before the next group comes along. This often makes treatment costs several times as high as they would be in a steady twenty tonnes-a-day operation. Hence another technical trap.

This problem has resulted in the survival of a crushing device described by Agricola as far back as the middle of the sixteenth century. Called the stamp battery, it was the only mill I knew as a child on my father's mine. Its great merit is that it will treat as little as a few tonnes a day in one unit; rocks and water go in under the thumping stamps and finely ground—well, fairly finely ground—particles get splashed out through a screen. The noise, a distinctive rolling three-beat or five-beat rhythm depending on the number of stamps in the battery, is considerable. In the early days of the Witwatersrand there were thousands of stamp batteries, creating an awesome thunder that could be heard for many miles.

Unfortunately for most small-scale miners, stamp mills are unavailable outside Southern Africa, although relics survive elsewhere. Deep in the remote Eastern Province of Zambia I once came across an abandoned mine (it was said that the owner had gone off to war in 1941 and never came back) where a small stamp mill had been left. Bush fires and termites had consumed the massive wooden posts that held it together, but the iron and steel components were lying in the long grass, barely rusted and ready for use nearly fifty years later.

So there are three traps, legal, financial and technical, waiting for African mining entrepreneurs to fall into. And of these the legal trap is the deepest. Just how virulent its effects are can be seen by comparing the make-up of the industry in a country where this trap does not exist—Zimbabwe—with the rest.

In 1994 the Zimbabwe Geological Survey published a major scholarly work entitled *Structural Controls of Gold Mineralization in the Zimbabwe Craton*.[92] It was written by two British expatriate geologists stationed in the Survey, S. D. Campbell and P. E. Pittfield, and in the introduction they drew on the production data for more than 700 mines to demonstrate a dramatic relationship between the numbers of mines and their output. Intuitively there are many more small mines than big mines, and so the distribution will be a highly skewed one.

However, when the cumulative number of mines were plotted on a log-log scale against their cumulative outputs (so many that have produced over a hundred kilogrammes, so many that have produced more than one tonne and so on) the distortion vanished. A very nearly straight line emerged, as the following graph shows. The only point that falls off it (slightly) is right at the top end. The plot suggests there should have been about five mines which produced over 100 tonnes; in fact there have been only two, the Cam and Motor and the Globe and Phoenix. Both have been effectively closed for thirty years, but perhaps there are other elephants lurking in Zimbabwe's geology and the indefatigable hunters for 'world class' deposits in Zimbabwe are right after all. Zimbabwe is not the only country to show this phenomenon. Western Australia, another country with archaean gold geology and a (until recently) liberal mining law, also gives a straight line.

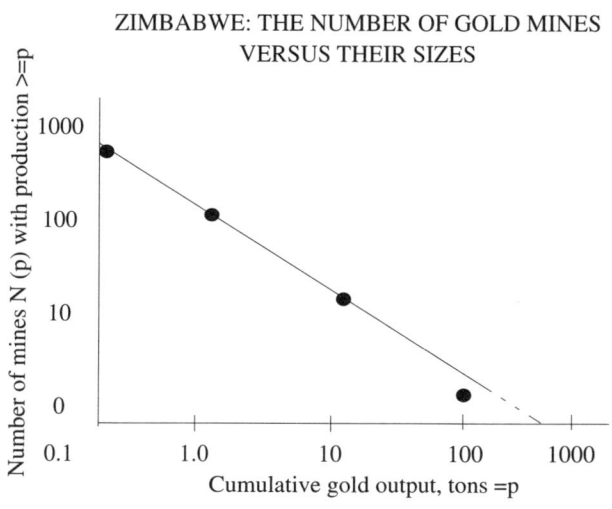

All poor together

What has happened is that in this environment the 'property right' nature of mining claims, which allows the owner to sell, option or tribute them without interference from the government, leads to a vigorous prospecting and claim-pegging culture, together with an equally vigorous marketplace in those claims. As a result the size and numbers of the mines that develop tends to reflect the natural distribution of geological opportunities.

It is interesting to extend the graph to include the very smallest mines. Assuming that the line can be extrapolated backwards to include those that have produced only ten kilogrammes or more of gold in their history to date, the total numbers run up to around the 3,000 mark, which is of the same order as the 4,000 mines thought to have existed in the country.

Countries that do not have a free-access, finders-keepers, full-property-right type of mining law end up either (as in Tanzania) with no mines apart from small-scale diggings, or (as in Ghana) nothing between the small-scale diggings and huge Ashanti-type operations. The graph that results is shown below.

Much has been spoken in development assistance circles about an alleged 'missing middle' in Africa. For example Africans either walk or ride in or on vehicles. They seldom cycle. It is

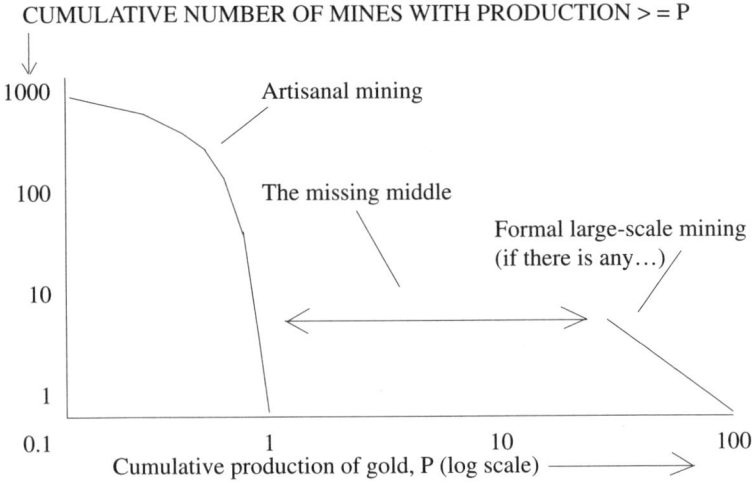

speculated that the vulnerability of bicycles to theft has something to do with it. Another, more likely argument, is that they have become too poor to afford bicycles any more.

Here, however, the missing middle is not only plain to see but its causes are manifest as well. What the diagrams also show is that there are plenty of small gold mines still with the potential for development into larger ones. That won't happen if changes of ownership become a matter of negotiation with government agencies. In Zimbabwe the existing system is becoming increasingly moribund, but that is a matter of inadequate departmental funding, not that it is a bad system. It is, from wide personal experience, the best in Africa.

• • • •

So there it is. Development assistance caused Africa's mining laws to be rewritten to be appropriate to 'the twentieth-century concept of mining'. The effect was to institutionalize corruption, to prevent the bottom-up development of an indigenous mining industry and to limit ordinary Africans who had found valuable minerals into exploiting them with their muscles alone. Heavens only knows what such well-meant aid interventions have done to other sectors of Africa's economy.

It has to be acknowledged that, despite this disastrous legislation, artisanal activity has produced some big miners, if not big mines. In 1994 I was in Mali. The day before I left Bamako for Zimbabwe I was asked by a government official to meet a local businessman, a quiet fellow in nondescript robes, said to be an artisanal mine owner, who wanted information on gold refining.

'Acheteur?' I asked, for surely a mere *minier artisanale* would have no need of a refinery.

'Non, il est minier,' I was assured.

Now I flatter myself that I know something about gold refining; had I not done the feasibility study on Zimbabwe's own refinery? So I took it on myself to tell this gentleman severely that a gold refinery was not economic if you had less than about twenty tonnes of gold to refine annually. Even if this amount was

All poor together

available, it costs so little to refine gold (a couple of dollars an ounce) in relation to the insurance and transport costs from a place like Mali, it would still hardly be worth it.

My French, which has caused even fierce Tuareg tribesmen to flinch, was probably not up to getting this last nuance over. Ah, said the miner, yes, but that was not the problem he had. He was already using a refining process; it had come from Switzerland and used lots of expensive chemicals. He was looking for a cheaper one.

Already using a refining process? Yes, he said. He produced a photograph. There he was, sitting on a chair—no, a throne. It was made of gold bars. At a rough estimate his perch was worth about fifteen million U.S. dollars.

This fortunate Malian was working eluvial gold in the Kenieba greenstone belt, on which a junior company, IAMGold of Toronto found a massive low grade deposit in 1993. IAMGold then searched for and found a mining partner, the biggest you could get, the Anglo American Corporation of South Africa. This is now the Sadiola mine which is currently producing over twelve tonnes a year—half of the output from Zimbabwe's hundreds of mines—from one big open pit operation. So maybe he would have needed a proper refinery after all.

Chapter Fourteen

IT'S GOOD TO TALK

It fucks you up, your donor's aid
They don't mean to, but they do
They pass on all the faults they made
And add some extra, just for you.
—With apologies to Philip Larkin (1922–1986).

Never be a taxpayer in an impoverished country. Relative scales might show that you are being taxed no worse than Sweden, say. The Swedes or whoever will certainly tell you this, implying that you don't know when you are well off. But the Swedes get back services for their money—good hospitals, well-equipped schools and a trustworthy police force. You will get none of those; there is no money for textbooks, barely enough for teachers, and hospitals are places you go to die in. As for the police ...

You are paying not just for incompetence, sloth and corruption, you are also paying for the unholy combination created by your government and aid donors. Civil servants dream up (or are urged into) projects. The money flows, experts arrive (very young, but then even doctors look young these days), the clinic, road, dam, whatever is built, the staff are recruited, trained, the ribbon is cut. Then the experts leave, their money stops and the government, it turns out, has not got enough of its own to pay for medicines, maintenance, pumps or whatever. The staff languish on modest salaries, sometimes months late, their skills fading away. The clinic is neglected, the road crumbles, the un-irrigated crops wilt. All that happens is that there is yet another small tap for salaries opened in the Treasury's budget, joining a million others. Combined they make a great flood, paying out taxpayer's money to a significant part of the population who are not actually able to do anything constructive, assuming that they were once competent (had the capacity, in aidspeak) to do so.

I was led to this chain of thought by seeing a reference in a

All poor together

technical journal to a United Nations Committee on New and Renewable Sources of Energy. This must be a slender survivor of all the urgent deliberations and declarations which ensued from the oil price crisis that accompanied the Iranian revolution, and presumably someone, somewhere, is footing the bill for this hardy growth.

I say this because the Ankara get-together that made such heavy weather of defining small mines was in fact my second United Nations' conference. In the first I had been very much an interloper at a very grand affair, held in Nairobi in 1981. It was about—and if my memory had not been jogged it would have taken some thought to recollect this, although it seemed significant enough at the time—New and Renewable Sources of Energy.

My excuse for going to that one—I was still in conventional employment then—was that at the time I was responsible for trying to get development started on Zimbabwe's second major coal mine, which boiled down to trying to find markets for the stuff. This was the coal that Bo Erikkson had come to find out about, so initiating the sequence of events that led to Buckreef and everything else. At the time it seemed important to find out what was going on in the camps of the rival energy sources. There were already mutters about global warming and despoiling the environment and so on.

New and renewable energy—solar power, windmills and so on—was pertinent then; the second oil price crisis in ten years was upon us and we in Zimbabwe, like a lot of people in Africa at the time, were labouring under petrol rationing. One of the 'new' sources of energy we were looking at then was oil from coal; hideously expensive but at least we had the raw material in abundance. In fact, though we could not have known it, we were about to enter an indefinite period of low oil prices and much of what was said at that gathering was already irrelevant.

A lot was said, of course. The size of the meeting rivalled that on the environment at Rio in 1992 or that of the women in Beijing in 1995. In the vast main auditorium at the Jomo Kenyatta Conference Centre distant figures ranted and gesticulated under the glare of television lights. The Palestinians declared that New and Renewable Sources of Energy were intimately bound up with their struggle against the Zionists, the representatives of the eastern bloc countries declared that N & RS of E were vital in defeating the capitalist onslaught, the western nations saw it (or maybe N & RS of E is a them) as the key to stopping the Soviet onslaught. But everyone was agreed on one thing; it was a vital weapon in the defeat of apartheid.

This nonsense went on for the best part of a week. From the very start of that period a committee worked day and night in an office in the tower atop the conference, preparing the final communiqué. When it finally came out it took several pages of achingly distended prose (all those issues and inputs) to say that N & RS of E was a Good Thing.

In the meantime there were a million urgent meetings in the halls and passages, a daily newspaper was produced that detailed the frenzied churning of words of the moment, a United Nations' stamp celebrated the event with a flower picture designed by Princess Grace of Monaco and a woman delegate was shot dead and her car hijacked.

I was an anomaly there. I worked for what I discovered was rather disdainfully called a transnational corporation and I was developing, or trying to get developed, a wicked old-fashioned non-renewable resource. So the accreditation people consigned me to a sort of counter-conference called the NGO Forum. Like plate tectonics, it now seems improbable that my first encounter with this abbreviation—I use it, hear it, daily—was so relatively recent. They were amazed that I didn't know what an NGO was (and I am now in turn amazed at your ignorance; it means non-governmental organization).

While this term might seem fairly all-embracing, in the strict sense an NGO is aid-speak for people who do things with money that doesn't come from governments (e.g., Oxfam, old version). However, because they have turned out to be relatively efficient in delivering particular types of aid to the deserving, at least in comparison to third-world administrations, they get a lot of western government money nowadays (e.g., Oxfam, new version). Indeed a vigorous little market place has developed, in which

All poor together

NGOs compete for the use of this money. It is probably the most valuable improvement in aid disbursement since the development assistance industry was invented.

Transnational corporations may seem to fit the general category of NGOs but they are not. They are, well, just not, you see.

Nonetheless, for the purpose of getting me out of the way, a blind eye was turned to my grubby antecedents and I went across to the Nairobi technical college to meet my fellow NGO-ers. They were an improbable collection for a conference on New and Renewable, etc. There was a fellow from the Save the Whales Campaign, a Colombian who had invented an improved village toilet, a representative or two of the Southern Baptist Convention and so on. A few—a very few—did have some link to N & Rs of E; there was a person there from the Intermediate Technology Development Group, for example, but for most the connection seemed tenuous at best.

Never mind. For here as well there was a chance to strut. Although there were no television lights, only the occasional cameraman, and no translation equipment, and the venue was a modest lecture theatre, here too, just like the mother of all conferences over the way, a final declaration was being hammered out by a fractious committee. However, as the revisions emerged they were debated in refreshing anarchy by the whole room instead of being the subject of thousands of little intense encounters in hallways and passages. At some stage, and with much preliminary excitement and false alarms, a rather patronizing message was brought down from the secretary-general of the conference (not the Secretary-General of the United Nations, you understand) to say how valuable the deliberations of the NGO Forum was, and how useful its input in the deliberations of the great and the good across the way.

Of course, whatever was finally crossed and dotted at the NGO gathering has slid into even greater obscurity, if that is possible, than the final communiqué of the main conference. Yet that glimpse of the UN at work was an invaluable piece of education into the ways of the wider world that we ex-Rhodesians had emerged into, stumbling and blinking.

It's good to talk

The United Nations was also present—I am not sure if they were not responsible as well—at a third conference on small mines, this time in Iran, in 1989. At that time I had as yet to become a (perceived) expert on small mines; acknowledgement came only later when. 'Ah, Hollaway! Small-scale!' was used in recognition of my singularity by more than one minister of mines on our first being introduced.

The route that led to this misplaced status is worth recounting, if only so that other aspiring experts might benefit. For many years I had had my name on the consultant rosters of both the World Bank and United Nations (and various others, such as the EU), but apart from one or two tiny jobs (such as demonstrating that Buckreef mine had consumed, not generated, wealth) I languished unsummoned.

An explanation was proffered in Tehran, of all places, in 1989. It happened like this. In 1987 there was a mining conference in Zimbabwe and I was detailed to lead a group of visitors around one of our larger mines. During this visit two smallish gentlemen with designer stubble and buttoned-up shirts kept getting under my feet, as they scribbled backwards in their notebooks. It emerged that they were from the Iranian geological survey. They would be grateful if I could show them some of Zimbabwe's smaller mines. So we had a look at a couple of little gold mines and some others mining things like feldspar and andalusite. During the trip it also emerged that they had their own potential gold mine in Iran.

Slowly a deal evolved. I would have my expenses paid to go to Iran to attend a conference in a few months' time (the Iran-Iraq war would be over by then, not a problem, they said). I would give a paper and talk to the state mining corporation about how to develop this mine.

So it was that, a year later than planned, I found myself in Tehran, waiting for several days in a once-plush hotel whose name I never discovered, but whose pre-revolution identity could be established from the massive neon letters lying where they had been thrown on the waste ground behind it. THTYA they said.

The war was still on, and there was the odd distant thud as missiles arrived in or near the city. The conference was accordingly

229

All poor together

being moved to Kerman, a remote city based on an oasis near the Afghanistan border. While waiting for this in Tehran I ventured out into the suburbs by bus, getting strange looks but generally friendly treatment, and swam in the hotel pool, mainly to avoid the incessant all-pervading, unchanging musak (remember that fellow with the pan pipes? The THTYA's only tape). The main users at that time were tubby senior military men, presumably called back from the front line to attend meetings. They seemed a little disconcerted by my presence in their midst.

Non-alcoholic beer featuring the taste of charred caramel enlivened the drinks menu. In the extensive ground floor lounge a massive tapestry hung around the walls. DEATH TO THE YANKEE SATAN it read.

In this sober, hushed, tootling environment I whiled away the hours swimming and working on the paper I was to give. I also read the improbable feasibility study on the Iranian gold prospect. It had been put out by the Czechoslovaks, those well-known gold-mining people, and was principally concerned with the elaborate arrangements for cyanide destruction. Everything else was pretty scanty. On television the mullahs orated endlessly before enthralled crowds.

I was the only, well yes, European in the place until a day before we were due to fly out to Kerman. At breakfast that morning was another whitey; we looked at each other and nodded distantly, as was proper. Clearly a fellow Anglo-Saxon. Of course in the end we ended up talking. He was here for the conference, from the United Nations, from that Mineral Resources Branch which has since been buried under a mound of initials. He was a nice sardonic Canadian, John Guy Bray, and I grasped the opportunity to ask why I had never heard from them, even though my name was on their consultant's roster. One of the moments when one contrives to emerge as gauche and naïve at the same time.

'Ah, the roster ...' he said, thoughtfully. 'Well, of course, we never really look at the roster.'

'No?'

'No. Basically we give the work to the last one who has come through the door.'

'A recent memory of a name, a face?'

'Just so.'

He wasn't joking. So every year since that time I have made a point of going to the offices of the appropriate bit of the United Nations in New York (not too difficult to find; their incessant renaming and reorganization may have imperilled their existence but the survivors' offices are in the same place). I do the same in Washington for the World Bank, which is not in decline, but for that reason its members are more difficult to find, being often away. On mission, as they say.

Before closing the Iranian drawer, my arrival and exit from the place is worth recalling. Getting in had been difficult, if not unusually harassing at the airport. The worst part was after having finally come into the arrival hall, if that was what you would call it in Farsi, there was nobody there to meet me. I sat on a bench amidst a great concourse of gowned Iranians feeling conspicuous and forlorn; a combination that tends to be the lot of solitary travellers in seriously foreign parts. A loudspeaker bellowed intermittently and incomprehensibly above my head. Eventually a large Iranian plonked himself in front of me and demanded 'You Japanese?'

I know that Iran was xenophobic and isolated, and so for that matter was Zimbabwe, but this fellow seemed to be making a meal of it. I told him where I was from, which, I suppose understandably, was doubly confusing. After a certain amount of continuing mutual incomprehension he pushed a paper at me. 'You?' To my surprise it had my name on it.

'Mining,' I said, pointing at myself.

'Yes, yes, you mining. We call you always, why you not come?' He was pointing up at the loudspeaker. 'Come now.' So I came.

That was easy enough. Now for getting out. During the conference a confused telex arrived in Kerman that had been transcribed into Farsi in Tehran. It said, as far as I could tell (the Minister of Mines, a mullah himself of course, had kindly translated it back for me), that I should return to Zimbabwe as soon as possible for a meeting with an important potential client.

All poor together

So I left the conference a day or so early, missing tours of various mines, and set out on a lone trip by air back to Tehran. In that battered city I met officials of the state mining corporation, who presented me with a small sack of ore from the mine-to-be. I commented temperately on the strange Czech proposal, but it quickly became clear that Zimbabwean advice, relevant as it might be, was not sought. What they wanted was help from South Africa, the big gold country.

And welcome. The gentle, or at least persevering, reader will understand why South Africa is not the most relevant model to follow for small gold mines, but it is normally futile to try to explain this to government mining officials. Even the Iranians would be certain that I was saying it only to ensure that I could return to work in their splendid country. Actually, even getting a visa to go there was gulp-making, for part of the formalities involved having to look at an elaborately bound set of vivid colour photographs of war casualties. This in the rather too close company of one of the lightly-bearded young fanatics that they seemed to have so many of. The South Africans could have this lot.

After the meeting I was dropped off at the airport. Or near it; a complex of barricades and tank traps prevented cars from getting too close. I had no reservations for any of the flights I was going to have to take, I had fifteen kilograms of gold ore in addition to my luggage, I couldn't speak the language. I staggered ahead.

Five hours later I was on a flight to Dubai. This was the second aircraft I had boarded; the first gave up after taxiing around for a bit, which was somewhat distressing in the light of the departure ordeal up to that point. This proceeded, as far as I remembered a day or so later when I thought to write it all down, along the following lines:

> The standard body and luggage x-ray at the entrance. The sack of ore was opened, heads shaken and then, to my relief, it was returned.
>
> A ticket check at the customs entrance (my original ticket was clearly unintelligible to them, but in the end they let me through).

232

It's good to talk

A customs search, which was very, very thorough. Everything was taken out and felt over, my wallet was emptied and the contents examined piece by piece. The small plastic sack of ore — a soil sample, I protested to anybody who would or could listen — was dug into, and x-rayed again.

A policeman who stood behind the check-in counter to, presumably, check on the check-in clerk, and who, indeed, took part in the complex negotiations that I had to undertake to get my ticket changed. Allah's will, not only did the clerk have a useful amount of English but there was a flight that afternoon to Dubai, and there was a seat. It gave me great pleasure to leave the ore sample with him.

After paying the departure tax (1500 Rials; no receipt was issued, I hoped it was all right) the next stop was at the entrance to the immigration area, where two policemen breathed heavily over my boarding pass, passport and ticket, and, to my slight alarm, took away the embarkation card I had filled in.

In the immigration area an official sat in front of a big battery of file card boxes and hunted and hunted and hunted for, I presume, my name. Eventually, visibly disappointed, he stamped my passport and let me proceed; I assume I was not on the lists of those who were not allowed out.

After immigration I was given a form to fill in for the departure tax I had already paid and this was then stamped.

Next I was put in a little room with four soldiers—well, I think they were soldiers; they certainly had Kalashnikovs handy—and was given, effectively, a strip search along with a further rummage through my wallet and luggage.

Finally out, thank the Christian God, into the departure area where my passport, ticket and boarding card were examined again and I was surprised to be given back the

All poor together

embarkation card—it was finally demanded on boarding the aircraft; that was the very last check.

In the event the little sack of ore went missing somewhere en route. I saw it come off the aircraft I arrived with in Addis Ababa, so its non-appearance wasn't due to Iranian paranoia, more likely the confusing of Harare in Zimbabwe with Harar in Ethiopia. Also the mysterious important client who had caused my unplanned departure turned out to be a completely non-mysterious nonentity. The English-Farsi-English translation had evidently caused the loss of some nuances. He had found another consultant by the time I arrived, anyway.

All this is by-the-by. The point was that I learnt in Iran that the United Nations is just the same as any other large bureaucracy, except more so. This means that there is a particularly complex system of rules and regulations—the need to have your name on the roster, for instance—that are used for defensive and protective purposes. Beneath this shield, business goes on in the ordinary way, with people you know helping you (to get jobs or whatever) and you helping them (by thanking them in your reports, by writing what they want, by not denouncing them as incompetent egocentrics and so on).

I was once told by a World Banker that there is a widespread impression there that UN consultancies are often awarded on a kick-back basis amongst what might be called affinity groups. I have to say that it never happened to me and a lot of the people I was dealing were, well, sort of from my own affinity crowd. Bureaucracy, yes, indecision, yes, lack of focus, certainly, lack of maturity, often, lack of ability, sure. But corruption, no, not to my knowledge.

But the poor devils—just think of having to work for the United Nations! Trying to be all things to all men, perpetually diplomatic, attentive, cordial, helpful even, to government officials. And so many of those, from any one of the hundred and eighty-five countries that are their masters. I don't blame the international bureaucrats for being, on occasion, pretty arrogant to the consultants and the like that they hire. They have to kick the cat, sometimes, like the rest of us.

Chapter Fifteen

WHAT A BUSINESS!

... You cannot, therefore, 'invest' in mining stocks; it is only possible to speculate in them.
—Doug Casey's *International Speculator*, June 23, 1998.

Mining has an ineluctable lure for a surprising number of otherwise sober and cautious individuals. Perhaps they are driven by the desire to announce at a dinner party (during, say, a discussion of the value of the gems on the rings worn by the hostess) that 'as it happens, I have a little mine in Africa'. It cannot be for the money that German doctors, Swedish dentists and American real estate executives send me off on hare-brained trips to God-forsaken places to look at some imagined Eldorado that they have been tempted to buy. It is bad enough for me, and I *am* doing it for the money. There just seems to be a level of disposable income which triggers folly.

Not necessarily on this sort of small scale, either. Such an occasion occurred in the 1970s, when the oil companies amassed unprecedented amounts of cash from the oil price hikes of that era. They looked around for ways of using this and thought they discerned a synergy between their business of pumping stuff out the ground, and that of digging stuff out of the ground. After all, geology was at the foundation of both activities, and they had lots of geological experience. In addition, although metal prices had tended to follow oil prices up, the financial markets (who may have known a thing or two) were underwhelmed by this unusual bit of good luck for the mining business. As a result the shares of such companies looked cheap, and for stockbrokers and analysts, besieged by oil majors who were running out of places to park their money, it must have seemed churlish, foolish even, to discourage them.

As a consequence, the ensuing decade was notable because of the lemming-like rush of the oil majors into this sector, waving their cheque books and stumbling over each other with their bids and counterbids. Their acquisitions were remarkably predictable; they seemed to have been purchased more for affinity than for profit. BP bought out the venerable British company Selection Trust, Royal Dutch Shell bought out the similarly august Dutch mining group Billiton, the Americans acquired big American miners (Atlantic Richfield bought Anaconda copper, Occidental Petroleum bought the Utah International Group, Sohio bought Kennecott Copper) and so on. Exxon went in at the deep end and acquired numerous exploration concessions. Almost all bought coal mines or uranium mines as these are also in the energy business, are they not? Therefore they would benefit from the next oil price shock. That there might not be another one in their lifetime did not enter anybody's thoughts.

The oil company's timing was terribly bad. After about 1982 oil prices languished and with them metal prices even more so. They lost heavily and started to back out, cutting their losses. The old-time mining houses came back into the game; Rio Tinto picked up BP's mines for U.S.$ 3.7 billion, the South African mining company Gencor acquired Billiton from Shell, BHP got Utah International, and thuslike.

By 1990 the only oil major with mining interests was Shell, with some coal mines, and it sold those off in 1999, at a time of exceptionally low coal prices. Nonetheless, this rump of their investments still fetched about a billion dollars, which gives a measure of the amount of money they had put into the sector. 'To run a coal business professionally, you need real coal people, and we found that this is not the sort that gas and power companies like us breed,'[93] said the (lady) executive who ran the appropriate Shell division at the time. Shell had first invested in coal something like twenty-five years before, so they took some time to find out about the breeding bit.

Very few individuals have become rich from actually mining for themselves. The success stories of the Californian forty-niners came from the storekeepers who supplied the diggers with their tools and rations. This holds true today, particularly at the bottom end of the scale, where people are 'diggers' as opposed to 'miners'.

What a business!

The most convincing evidence of this comes from an unlikely source. In the late 1930s Dr Robert Mackay was in the Tanganyika (now Tanzania) Mines Department, responsible for an area in the south of the country which included the Lupa goldfield. In 1948 he wrote a brief account of his experiences as a mines inspector there at that time, which is invaluable despite, or perhaps because of, its stunning political incorrectness.[94] The situation in Tanganyika before the war was that there was no income tax to evade, just a 5% royalty, and no flimsy local currency to shun (the East African shilling was linked to sterling). As a result, perhaps uniquely nowadays, there was no black market in the gold produced. Or was there?

The Lupa is in an area of high, rolling, lightly-treed uplands similar to parts of the Zimbabwe highveld, but cut by dramatic river gorges. The first gold to be found and exploited were alluvial deposits, discovered in the Lupa River by a William Cummings in 1923. In the way of things the find attracted people described by Dr Mackay of 'an adventurous and unconventional disposition'. Tanganyika did not have what was then known as a 'colour bar', and, to quote:

> These men and those that followed them differed widely in race, background, education and past occupations. Senior army officers and ex-cabinet makers, retired government officials and barbers, erstwhile planters and professional hunters, traders, artisans, prospectors and miners of British, German, Greek, Indian, African and many other nationalities made up the community.

By 1936 there were some 770 'non-natives' working on the alluvials, employing between 15,000 and 17,000 local labourers. This was an indubitable gold rush; it was probably the greatest concentration of 'non-natives' and certainly that of 'industrial workers' in Tanganyika. By then attention was starting to move from alluvial gold to the eluvial gold around gold reefs in the uplands, whence the Lupa River and its tributaries rose. A few big eluvial nuggets were found, perhaps from reprecipitated gold as a consequence of secondary enrichment, but then the 'mother lode' quartz reefs were discovered. These produced most of the

gold in the long run, sustaining formal output until the late 1950s.
The lengthy dry season meant that those working the gold in the riverbeds had to develop a technique that did not use water for concentrating it. These 'dry blowers' used a fan to separate the heavier gold from the lighter river sand. They were noisy and inefficient, were marked by a great drifting cloud of dust and probably led to much unseen damage by causing silicosis in the lungs of the operators.

The efficiency of this gravity concentration method, for that is what it was, at the Lupa was very low.* Nonetheless it worked, and forty years later a Russian aid programme (judging by Soviet 'assistance' elsewhere in Africa this may be a generous description) had massive mechanized Soviet blowers reworking the river sands in conjunction with our old acquaintance, the Tanzanian State Mining Corporation, Stamico. According to some log sheets I saw the recovered grades were about 160 mg of gold per cubic metre treated, which would not have been profitable even in a country with good infrastructure and services. A rule of thumb for profitability for alluvials is about twice that.

So who made money from the Lupa alluvials? Not the Russians, certainly. But back in the thirties?

Dr Mackay's analysis of the figures produced some unexpected results. Very few fortunes were made, and these were the result of spectacular but short-lived finds. Rather the pattern was surprisingly consistent: ten of the names of the most productive groups of alluvial syndicates had been ranked amongst the top twelve of previous years. As he said—

> The ability of certain persons to head the list in consecutive years, despite changing ground conditions, seems to

---

* [Disregard this if you have no liking for maths.] Separation processes are controlled by a formula $(S_H - R)/(S_L - R)$, where S_H is the density of the heavy material—in this case gold of about 19 grams per cubic centimetre—and S_L is the density of the light material, in this case sand, of about 2.5 grams per cubic centimetre. R is the density of the medium in which the process is occurring. The higher the result, the better the separation. Water is the usual medium, with a density of one, giving a separation ratio of about 12. If air (with a density of about 0.0013) is used, as on the Lupa goldfields, the separation ratio becomes effectively S_H/S_L, or about 7.5.

rule out luck as the main factor. Neither was energy by any means the sole road to success. The answer seemed to lie in organizing ability or an indefinable 'flair', or more often a combination of these factors.

He goes further. He examined the output by national and racial groups. The following (for today) hugely embarrassing table ensued:

Nationality	Output Ounces/Month
British	11.2
South African	10.9
German	7.5
Miscellaneous*	6.6
Greek	6.1
Goan	4.0
Asiatic	3.0
Native	2.0

* The nationals of some dozen countries, mostly European and middle-eastern.

But wait. This was not an exercise in racial one-upmanship. Quite the opposite. Dr Mackay's explanation was a deeper one; he argued that this list approximated to the standard of living for the respective groups. The playing field was, after all, fairly level; the claims were open to all, were small (300 feet by 100 feet), changed hands often, and picks and shovels were cheap. Indeed, he implied, once a claim had been acquired it was possible to get things going on credit alone (his phraseology is 'capital was not a great necessity at a time when labour could be engaged speculatively'). His conclusion is that the claim holders must have been working—or more correctly working their labour—for what they judged they needed, not what they could aim for.

His figures again savaged the persistent myth that gold mining was a good way to make money. Rather the opposite; in 1936 out of 937 claim holders, 512 lost money while all the rest broke even or made only a modest profit. Only 15 earned a surplus of over £100 a month in that year, equivalent to about £800 for the average operating season.

As Dr Mackay said, 'The immediate question which presented itself was how losses of this sort were supported ...' His conclusion, which was derived, presumably, from practical experience, was that those cabinet makers, barbers and so on amongst the losers survived by providing services for the successful diggers. One hopes for their sakes that the adventurous senior army officers and retired government officials were amongst those few successful enough to live off their gold winnings.

A question mark remains. Was all the gold really being declared? Did not some people have an incentive to smuggle some out? Those Germans, for example. Why was the official output of such vigorous task-masters relatively low? In common with all the ex-German colonies at the time there was strong covert support for the Nazis, coupled with a marked antipathy to paying any form of taxes or royalty to the victors of the First World War. But was gold being accumulated secretly to buy weapons and ammunition? As Dr Mackay noted, 'the relative magnitude of the Germans' winnings is somewhat of an anomaly unless their political preoccupation [it was 1936, remember] or a feeling of insecurity could account for it.' They were, almost certainly, keeping the gold for their own use. Read Roald Dahl's second volume of autobiography.[95] He, as a junior Shell Oil official (and reserve officer) was engaged in a dramatic confrontation with several hundred German ex-colonialists heading south from Dar-es-Salaam on their way to (neutral) Mozambique when war broke out three years later.

And the Asians? Gold has been the only sure source of wealth to them for centuries, and still is. The temptation to put aside some for their daughters' dowries must have been huge; India is the world's largest single gold importer, about 400 tonnes a year nowadays, much of it smuggled. Africans? Here the opposite applies. African wives-to-be don't bring dowries, they are purchased from their fathers. This difference (whose implications are much more profound than it might seem) means that the need to accumulate the wherewithal to buy a wife will loom large in a young worker's thoughts. The easiest way to accumulate wealth would be to steal a little.

This tends to be confirmed by a description of the system of worker remuneration in the Lupa in the 1950s.[96] By then most of the alluvial gold had gone, as had the 'non-native' claim holders, and the forerunners of today's artisanal workers were active on reef discoveries. It was recorded that African claim holders did not pay their labourers but worked on the 'prizi' or prize system of rewards for what was delivered. This, not unnaturally, evolved into a system of labourers holding on to the gold they recovered until they needed cash. Then they would sell it elsewhere, usually to another claim holder. In the meantime the first claim holder in his turn was being sold gold brought to him by labourers from other mines. The system was known as *'pata mali'*, Swahili for 'get cash'.

> Every year saw a couple of delegations of claim holders arriving at the Mines Inspector's office to ask him was he was doing to stop the pata mali system. The short answer to that was 'Nothing', as the remedy lay in the hands of the diggers themselves; suggestions that they should not buy stolen gold were received with looks of astonished incredulity.[97]

By now an income tax system was in place, and evasion was said to be rife. The same author—another mines inspector—notes that, based on his own knowledge and experience, some mines were simply fronts for illicit gold buying.

It seems likely that when these various compensating factors are taken into account, the discrepancy between the productivity of the various nationalities and races operating alluvial claims in the Lupa may not have been all that great after all. Was the data that Dr Mackay so carefully recorded more a ranking of honesty than of management?

The Lupa, he said wistfully, may be the last of the old-fashioned gold diggings, 'now that we have reached an age of planned economics and attendant regulations.' Perhaps so, but as readers know, a lot of new-fashioned diggings have started up since he wrote that in 1948, including many in the Lupa goldfield, where reef discoveries by artisanal miners have continued to proliferate.

By 1996 there were no almost no formal mines at all in Tanzania, but local people were more active in mining than ever before. A World Bank artisanal mining baseline survey team visited 105 sites (38 gold, 36 gemstones, 31 other minerals), and these were only a part of the total of artisanal workings.[98] Some things do not change—

> Claim holders and pit financiers don't pay salaries to mine workers, but they extend various incentives which are aimed mainly to retain miners to work in their pits.

These incentives included compensation for the cost of food and medical treatment and providing food on credit. The 'prizi' and 'pata mali' systems are not dead yet. It's still very hard to make money from a gold mine.

Chapter Sixteen

THE BARBARIC RELIC

Not many noticed what would have once have been a heroic act. That was at least partly because heroic language was avoided. Men did not speak of the final abandonment of the gold standard. Instead it was said that the gold window was closed. No one could get much excited about the closing of a window.
—J. K. Galbraith, *Money. Whence it Came, Where it Went*, 1975.

The most common question asked in Africa when you cautiously admit to being in mining is not a languid 'And what do you do there, exactly?' but an intense 'Now tell me, what is going to happen to the gold price?'. This chapter will give the straightest answer possible. And it is not 'I don't know.'

At the end of 1998 the total amount of gold in whatever form present above ground was thought to be 137,400 tonnes, or about 4.4 billion ounces.[99] Pure gold has a density of 19.3, so this amount represents a fairly modest volume, specifically a cube with sides of about 20 metres. At that time the value of this big yellow lump (the gold price was of the order of U.S.$280 an ounce) would have been about one and a quarter trillion U.S. dollars.

Value and density makes gold a handy asset to move around. Perhaps the most dramatic instance of this was in 1968. It was a period when faith in America's ability to contain the communist empire had started to fade (the Vietnam war was already beginning to appear unwinnable) and with it so did confidence in the formerly almighty dollar. The inflation rate of this currency, which had been ticking along at about 2% ever since the Second World War, had leapt to ten times that much.

Until then the non-communist world was ordered financially by the Bretton Woods agreement, which had, as well as creating the World Bank and the IMF, set up a system of fixed exchange

All poor together

rates all linked to a set price for gold of U.S.$35 an ounce. This number had been established all the way back in 1934 as part of Roosevelt's 'New Deal', by a process which does not bear close examination. It originated from a graph which showed that industrial and agricultural production went up with the amount of gold in the world's economies.[100] The causality, as economists would have it, went from left to right; more gold, more production. More production, more wealth. In the heart of the world's greatest depression this seemed an interesting idea. But how to get more gold, quickly?

The answer was part pragmatic, part smoke and mirrors. The pragmatic part was to raise the gold price; eventually the mines would respond. But the magic part was that the effect would signal ahead: by raising the price of gold in dollars, each dollar would represent less gold, and since gold featured heavily in the reserves, there would be more dollars available for bankers to lend. Controlled inflation. So, starting in late 1933, when the price was about U.S.$20 an ounce, the U.S. government began buying gold at prices which had increased to U.S.$35 an ounce in January 1934.

The price-setting system was extremely basic. The daily rate that the U.S. government would offer for gold was set at breakfast each morning in discussions between President Roosevelt, the Secretary of the Treasury and the head of the Reconstruction Finance Corporation (which because of a dereliction of duty by the Federal Reserve Bank had become the *de facto* lender of last resort in America). There is no reason to believe that their guesses about what the price should be were any better than anybody else's.

Unfortunately, as a stimulant to economic growth the scheme did not work. It seems that the president and the rest had forgotten that earlier in 1933 they had suspended the gold standard, so that U.S. currency notes could no longer be exchanged for gold. It was therefore immaterial to the bankers that there was nominally less of the stuff to back the currency bills they were handling: there was no increase in the amount of cash they had available to lend out. The mines responded, though, quite briskly. The amount of gold produced climbed from 793 tonnes in 1933 to 1,029 tonnes in 1936.

The barbaric relic

To be nit-pickety, the breakfast club's final guess of U.S.$35 an ounce was set as the central banks' buying price; thereafter if you wanted to buy from them (which after the Second World War relatively few people in the West were allowed to do; many countries still banned private gold holding) then you paid U.S.$35.20 an ounce. In any event this fixed price was not high enough to stop my father's minuscule gold operation (an adit and a three-stamp mill) from going under thirty years later, along with many other mines in Zimbabwe and elsewhere.

These little tragedies did not show up in the grand total of world gold production, for that was the period of the opening up of the Free State Goldfields. Between 1955 and 1970 South African gold output more than doubled, pushing the global figure from 1,000 tonnes to 1,500 tonnes.

But now, in 1968, the once omnipotent dollar was suffering a 20% inflation rate; in theory gold was devaluing by 20% annually. Was that really so?

Many people—especially those with significant disposable income—decided not. The lucky ones who were unhampered by exchange control regulations—exporters lumbered with what were known as Eurodollars, wealthy Middle-Easterners and the like—judged that, logically, the U.S. dollar was overvalued and gold undervalued and acted accordingly, and rapidly.

As the astute and the wealthy offloaded U.S. dollars there came a demand for gold the like of which had never been seen before. It is generally assumed that all that happens when gold changes ownership is that the bullion sits in the same place, in the same vault, but with a different name on it. Not in this case; the gold that was being bought was in Fort Knox. The purchasers wanted it out of there and in somewhere safer, like, surprisingly enough, the Bank of England.

Over 3,000 tonnes of gold made that particular journey in early 1968, 1,000 tonnes of it between the 8th and the 14th of March. Huge U.S. Air Force transport aircraft, whose normal task was to carry tanks and the like about the skies, had to be pressed into service.

From a miner's point of view, this panic did a number of useful

All poor together

things. For one, it gave the United States government a great fright. If gold reserves could melt away so easily then better the government not sell them at all.* This, handily, has kept about 8,000 tonnes, or 25% of all official stocks, off the market ever since that time. Also taking note were the central banks of France, Germany and Italy, who have generally followed the same policy. Collectively these four countries hold more than half the gold stored in the vaults of the central banks.

The panic also caused a number of modestly rich people who were shortly going to become enormously rich—the sheikhs of the Gulf States—to sense the usefulness of gold as a 'store of value'. You may remember that Dr Kamel in the middle 1980s got his Dar Tadine Tanzania started on the back of the continuing desire of Saudis and Kuwaitis to get out of dollars and into gold.

But best of all for the industry, it destroyed the gentlemanly arrangements of the Bretton Woods exchange rate agreements. Of course, in true central bank style, up to the very last moment all the officials concerned declared that they would not change the arrangements, that Fort Knox would be emptied before they retreated from the U.S.$35 an ounce price. But on 15 March 1968 the British abruptly declared an emergency Bank Holiday and closed the London bullion market.

There were frantic transatlantic telephone calls, the telexes chattered incessantly and then it was all agreed. The London market reopened with a two-tier arrangement; the central banks would continue to deal with each other at U.S.$35 an ounce, but they would not sell to anybody else at that price. The rest would have to take their chance in the bullion markets. In August 1971 this cosy arrangement ceased as well. The United States announced that it would supply gold at the full market price to other central banks. President Nixon, heroically if you like, had closed the gold window. Thereupon the gold price started on an eleven-year

---

* Not without one last try, however. In 1979, as the gold price soared, the United States and the International Monetary Fund jointly tried to stop this devaluation of the US dollar by auctioning gold. This failed dramatically; after 535 tonnes had gone they had to stop as the demand was so great that the price continued to soar regardless.

climb, culminating on the 21 January 1980 when it hit U.S.$850 an ounce.

After that it went wobbling gently up and down, and down, and down. That was the period when I was saying I didn't know what the gold price was going to do.

One reason was the obvious one, that gold now had a true market just like the other commodities, such as copper, coffee and pork bellies. This meant that it had a great, theoretically infinite, range of notional future values as well as a present price. So gold could now be sold forward in one way or the other, usually by hedging, by gold loans and by gold options. Hedging is accomplished by getting a buyer to agree to purchase a mine's output in the future at a price above the current one. Gold loans (actually leases) involve a bank lending gold to a project to finance it on condition that it receives the same amount of gold back at the end of the term period. The interest rate on this transaction is very small, and the scheme was therefore attractive in the inflationary period of the early 1980s, less so in the sober nineties. Options are another form of forward sales, more flexible than hedging, but, as some companies belatedly realized, much more dangerous in a volatile market.

While these financial instruments represented a splendid way for mines to hold off the effects of a falling gold price, they had a sting in their tail. One was immediate; the 500 tonnes or so[101] of gold that was at stake in these instruments at any one time was a threat to the market price itself, and tended to depress it. The second was more hypothetical; it was dependent on the gold price not going up suddenly, which, after all, seemed most unlikely. Even the Gulf War had not stirred it from its languor.

In mid-1999 I was asked to represent Zimbabwe at a meeting in Tanzania of the Chambers of Mines of Southern Africa. Arriving at midnight in Dar-es-Salaam after a 48-hour journey from Ouagadougou (via, if it is of interest, Niamey, Ndjamena, Khartoum, Addis Ababa and Nairobi) I was feeling somewhat frazzled. This was not helped by my taxi driver, who followed the hallowed Tanzanian custom of not even slowing his decomposing Peugeot 404 for pink traffic lights. Pink lights are red lights when

All poor together

nothing else appears to be coming through.

Consequently, when I finally got to the hotel (which was, as it happened, overbooked), there was difficulty in empathizing fully with the evident fury and excitement of my colleagues from other chambers, particulary the South Africans. It appeared that the International Monetary Fund was going to offload its gold reserves to raise money to help poor countries, thus further depressing the price below its current 20-year low. I cannot pretend being overwhelmed at the news; all I was excited about was getting a good night's sleep, somewhere, anywhere.

The meeting was to be held next morning in Arusha, under the snows of Kilimanjaro, and I got up again after about three hours' sleep in the manager's bed (no, he was not in it) to be driven in the bleak small hours back through all the pink lights to the airport.

In the half-light there figures moved sluggishly about the elderly 737 of Air Tanzania that we were to use. Ever since I had first climbed on board one, seventeen years before, the airline has had just two such aircraft, one called Serengeti and the other Kilimanjaro. They got around; shortly before that first momentous journey to Buckreef the Serengeti had been hijacked while on a flight from Mwanza, ending up at Stansted in the United Kingdom. Were they really the same aircraft still? Did I really want to know?

On the trip the IMF sale details were explained to me. The institution had over a hundred million ounces of gold reserves—that's about 3,000 tonnes—and was proposing to sell between 5 and 10 million ounces (155 to 310 tonnes) in order to provide funds to reduce the debt of the most deserving countries—those Heavily Indebted Poor Countries (HIPC) last heard of in Chapter 12. The snag was that, in terms of the IMF Articles of Agreement, it could not weaken its asset base. As a result only the interest on the money these sales would realize could be used for debt relief.

The back-of-an-envelope figures were something like this. The Chamber of Mines of South Africa estimated that a gold sale of five million ounces would depress the price by between U.S.$10 to U.S.$15 an ounce. This would bring the then price down to about U.S.$245 an ounce, and the total revenue from all those ounces would be of the order of 1.2 billion dollars. Interest

The barbaric relic

on this in blue-chip investments at 5% would be about 60 million dollars a year, so that would be the amount available to offset loans.

However, argued the South Africans, the 41 HIPC countries produce about 165 tonnes of gold annually between them (those are just the official figures; as readers will now understand there are very few developing countries where some gold is not panned out of the rivers or scraped out of pits). Taking the then dreadful spot price of U.S.$255 an ounce at that moment and taking another U.S.$10 off it would reduce their income from 1.344 billion to 1.292 billion—a drop of U.S.$ 52 million. This would leave a net gain of about 10 million dollars from the IMF's sales to set against the HIPC's debts of about 30 billion, or about one dollar for every three thousand owing.

I was in no mood for sober reflection on this argument; after all Zimbabwe was a 25-tonne-a-year gold producer herself and anyway my mental processes were clogged with jet lag. So that evening, after our own meeting, a-lobbying we all went. For at the same hotel in Arusha were assembled the mining ministers and/ or their senior officials of the fourteen countries of the regional grouping known as the Southern African Development Community (SADC). Indeed, our gathering had been arranged to coincide with theirs.

The SADC was founded in 1980 to represent the 'frontline states' in the struggle against Apartheid South Africa (when this phrase was not used, it was 'Racist South Africa'; this was the official name for the country on the Zimbabwe Broadcasting Corporation's news for a time in the 1980s, including the weather reports). Not incidentally, it also served as a means of mobilizing aid money to support the member countries. In this it had the whole-hearted support of the 'development community'. For here, at last, was a way for them to spend money on projects that would enjoy the support of just about everyone.

Apartheid was the ultimate knee-jerk word. To be 'for' it was worse than being 'for' poverty; the frontline states were not just combating this evil but were full of ideas on how to make their countries independent of South Africa. Every project that every

All poor together

SADC government had ever thought of was dusted off, given a regional spin and presented for funding.

Contributing to the excitement was the competitive environment of the time in the matter of how much aid a country gave. The Scandinavians had a declared objective of 1% of GDP, and regular league tables were published to keep up the pressure on the laggards (these were principally the United States and the United Kingdom; neither Ronald Reagan nor Margaret Thatcher were terribly in the spirit of the thing).

In its practical implementation, individual countries were given sectoral responsibilities, such as agriculture (Zimbabwe), infrastructure (Mozambique) and energy (Angola), and co-ordinating units were set up in the relevant capital cities, paid for by the host country and overseen by a co-ordinator who was a national of that country. These proved invaluable sumps into which development assistance could be poured, thus contributing usefully to the striving of countries to boost their ratings in the aid tables. Between 1980 and 1990 about U.S.$2 billion was given to the various SADC sectoral units. Some of this, like the money spent on improving Mozambique's ports, probably did some good.

Mining was a latecomer to this arrangement, but in 1984 the Mining Sector Co-ordinating Unit (MCU) was inaugurated in Lusaka, Zambia. This country was selected as its fading copper mines were still the biggest in the region. In common with the other units, the *modus vivendi* of the MCU was to put forward projects that would attract both ministerial endorsement and donor support, and in this—the support, not the projects—it has been broadly successful. For example, the unit had thirty-five projects under implementation in 1999, with about eight a year being completed. The cost to aid donors over the previous fifteen years for such MCU projects was certainly not less than U.S.$50 million.

Unfortunately, these have not made any detectable impact on the general stagnation of mineral activity in the region, a criticism that applies to many of the other SADC sectors as well. This carping ill becomes me, as I have been part of a SADC project dealing with the environmental effects of mining. Yet it is hard to think of any of the projects that have done more than gather data and voice opinions; I cannot bring to mind any significant mining policy changes amongst the SADC governments that originated from all this expenditure.

With the advent of democracy in South African in 1994, the SADC objective was achieved. However, with so many healthy institutional arrangements in place it was perhaps unrealistic to expect the grouping to proclaim victory and disband. Instead, South Africa became a member, and a vigorous one at that. Then, and perhaps finally, the post-Mobutu Democratic Republic of the Congo (DRC) joined as well. The decision to accept this vast piece of old chaos (apart from the Seychelles and Mauritius the only French-speaking member, and one with almost no democratic credentials) was doubly foolish. No useful extra aid money could be extracted by it; there were no portfolios left for it to adopt. In addition, its presence bid fair to split the grouping, with South Africa leading a faction which had severe reservations about admitting the country. Grotesquely this resulted in a situation where three members (Zimbabwe, Namibia and Angola) were giving military assistance to Mobutu's successor, Laurent Kabila, while the rest were at best neutral, and in some cases (South Africa and Tanzania were named) said to be providing covert support to a rebellion against him.

This then was the background to the mining ministers' annual meeting in Arusha; they were there essentially to rubber-stamp a fresh wish-list of projects from the Mining Co-ordinating Unit in Lusaka, which would then be used to line up aid donors for support. As such it was an ideal venue to lobbying for opposition to the IMF gold sales.

I was desperate for sleep that evening after a full day of meetings, but I did my duty with the others at cocktail parties, the formal dinner and in the lounges and bars of the hotel until the small hours. The process essentially involved adopting a grave mien and ensuring that one's name badge was the right way up. After that it was sheer *chutzpah*, made easier by the mist of fatigue I was living in.

All poor together

'Ah, minister, how nice to meet you at last; the chamber of mines has followed your work with great interest.'

Minister (peering short-sightedly at the speaker's name tag): 'Well, Mr, Mr ... er, (gives up) er, well, thank you.'

'Minister, I am sure you are as worried as all the chambers of mines are about the IMF gold sales.'

'Gold sales? Ah, yes, the gold sales ...?' (He is looking urgently around now for his permanent secretary, but that person has already been buttonholed by another Chamberling).

'Yes, we expect the gold price to drop by another U.S.$15 an ounce if they go ahead; exploration companies are already pulling out. Projects are being postponed. Jobs are being lost.'

That did it; anything that loses jobs is a Very Bad Thing to a politician. The minister's attention has been gained.

This gamesmanship seemed to have its effect. The SADC mining ministers' meeting was opened by the Tanzanian vice president. Tanzania is a prominent member of the HIPC class of country, and the key words of his speech came at the end. After reviewing the IMF scheme, he concluded as follows:

> I wish therefore to call on the IMF and leading industrialized countries from implementing the proposal, since that [gold] sale will depress further the prices of precious metals and deprive our countries of the much needed revenue from mining production.

Following the speech, a number of ministers got up to declare their solidarity. It was enough. The IMF could hardly ignore such sentiments coming from the inheritors of Julius Nyerere's moral legacy in the third world.

Of course, true to form, the IMF insisted that it was proceeding with the idea anyway. Indeed at a meeting that June in Cologne of the Group of Seven (these are the biggies: the United States, Germany, Japan and so on) the plan had received a resounding endorsement from these heavyweights. It appeared a done deal. In the meantime the Bank of England and the Swiss National Bank announced plans to sell off about half their respective reserves. Clearly the flight from gold was broadening.

The barbaric relic

But two months later the IMF abruptly abandoned its plan. In its place it came up with a convoluted scheme that would generate money to offset the HIPC debts by effectively revaluing some of its gold stocks (still on its books at only U.S.$46 an ounce). For the gold mining industry in the SADC region, which included South Africa's 470 tonnes a year, the Lusaka mining unit had finally justified those ineffectual millions spent on it.

Of course, there were other powerful voices raised against the original scheme—the World Gold Council for one—but I like to think that our lobbying achieved what was wanted. The very people, the Africans, the poorest of the poor, who were supposed to have benefited from the IMF gold sales, had publicly voted against it and so, after a decent delay, the idea was dropped.

And not just dropped. In late September 1999, Wim Duisenberg, the head of the European Central Bank, announced, on behalf of his bank and those of eleven of the Euro-zone countries, together with the UK, Sweden and Switzerland, a moratorium on gold loans and a maximum of 2,000 tonnes to be sold over the next five years. You will remember that the Federal Reserve Bank has always said that it will not sell gold, so the overall effect was that about 23,000 tonnes of gold would remain for at least five years out of the market. The gold price surged, and then the law of unintended consequences came into effect.

To return to selling gold forward and the business of options. Until this sudden jump, the gold miners had capped previous price rallies by seizing the opportunity to take advantage of the increase to lease gold from the central banks and sell it at the higher forward prices available. The banks would then be refunded in gold again at a later stage plus a small interest payment, at which point the mine has the option of restarting the process by borrowing more gold. Such is the attraction of these arrangements that in 1997 on the London Bullion Market about 450 million ounces of gold, or well over 12,000 tonnes, were traded—more than four times the amount of gold that is used worldwide in jewellery annually.

Now the European banks had stopped lending their gold out, so the market had tightened and prices had risen. A number of mining companies with obligations to repay in gold were unable to discharge their loans unless they bought the metal from other banks at the new high price and at high interest rates. For a company like Ashanti Goldfields, Africa's largest mining group outside South Africa, the 11 million ounces of gold—over 353 tonnes—that it had committed to 2013 now represented obligations, which instantly turned the profit and loss account negative by about U.S.$450 million. Abruptly it was in deep financial waters, and these, as always, were heavily populated by sharks. Other big mining groups, including Cambior of Canada and Newcrest of Australia, had the same experience.

These substantial hiccups notwithstanding, it is with a degree of modest pride that I record that the London *Mining Journal*, whose proximity to the City and distinguished antecedents has made it a primary source of financial information for the industry, said of the European bank's *volte face* that the decision 'appears to be a response to strong political pressure, designed to reverse the damage done to the economies of gold-producing countries by official sales.' In Arusha, it seems, we had built better than we knew.

Yet should they—the South Africans—have fought the IMF sales anyway? This may seem startling, but have a look at the figures. The central banks and the IMF hold between them something around 32,000 tonnes of gold, or about seven year's demand. This holding has always hung like the sword of Damocles over the gold price, more threatening when the price is dropping than when it is rising. For three years, from 1996 to 1999, the price dropped steadily, and increasing amounts of central bank gold was offloaded on to the market. During that time five central banks (Australia, Canada, Belgium, the Netherlands and Argentina) openly sold significant amounts of gold and a number of others did so more discreetly. However, there are about 125 central banks in total, and some of those—Poland, the Philippines and (surprise!) Russia were identified—were active or net buyers during this period. Belgium's sale of 300 tonnes in 1998 was all bought by other central banks.[102]

Between 1980 and 1988 this selling and buying meant that central bank stocks dropped by only about 2,000 tonnes. We have

The barbaric relic

already heard, for what it is worth, that, regardless of the European Central Bank's decision, Germany, Italy and France along with the United States had no plans to sell their gold anyway. At a guess, therefore, of that 32,000 tonnes, only 10,000 tonnes are going to be available for the marketplace.

However, the supply/demand pattern for gold is right out of kilter. Both have been growing steadily for many years, but by the 1990s about 2,500 tonnes a year were mined while about 3,500 tonnes a year were 'consumed', much of the difference having been made up by central bank sales. In practice, of course, gold does not get consumed, but is mainly used in fabricated items. Most of this is jewellery; in an average year this sector has an offtake of about 3,000 tonnes alone although much of this is from recycled gold. This jewellery itself represents an overhang on the market; the South East Asian crash of 1997 saw governments asking for, and getting, their citizens to hand over their gold jewellery to support the economy, leading to an unprecedented supply of 'scrap' in 1998.

Nonetheless, as personal wealth grows much of the gold mined is taken out of circulation, at a rate of very roughly 1,000 tonnes a year. So, again in terms of the crudest numbers, after about ten years there might be no more gold forthcoming from the central banks, and the demand for gold will far outstrip its availability at today's prices.

But there are three other factors to consider. South Africa, with its 25% share of the market, has slimmed down and restructured its industry to keep in business, but for thirty years it has been facing inexorable increases in output costs per ounce due to falling grades and greater depths. Only the periodic devaluation of the South African currency, the rand, has allowed the mines to survive. In the 30 years after 1970 output halved.

Elsewhere most of the new production since 1980 has come from massive open pit operations. These usually have a relatively short life—10 to 20 years typically—and many of them now have nearly exhausted their reserves. No doubt some will start to develop the underground extensions of their orebodies, where they exist, but this is expensive and often metallurgically difficult—

All poor together

look at Buckreef. Although new open cast mines are planned, it appears that overall the supply of gold will not grow, and may well shrink, despite a rise in price.

Finally, and most important of all, are the psychological factors. The gold market will anticipate the forthcoming physical shortage. Long before the central bank sales taper off, the price will start climbing. And when this starts, a further factor will exacerbate the supply situation. This is the political difficulty of selling a nation's gold reserves in a rising market. It was easy enough to offload when the price was sinking; there was almost a rush to leave this foundering vessel. After all, had not the sainted Keynes said that gold was a barbaric relic? Yet to sell off the country's gold reserves when the price is moving upwards? Would the nation ever forgive them? For central bankers, prudence is a core virtue. They are also very aware of the market place balance numbers above. Better to wait.

So there you have it; the price will go up, it will go up fast, it will go up a long way, it will go up for good, and it will happen in a very few year's time. The only cloud on the horizon is the faint possibility that a new Witwatersrand will be discovered, perhaps under the sands of Botswana, or that a cheap way is found to recover the tiny amounts present in sea water. Otherwise the views of Keynes, sanctified or not, will be ignored, and nations will continue to sit on large stocks of gold, just in case. ...

To come back to all that lobbying in the Novotel Arusha against the IMF plan and the question of was it in fact necessary? You now have the facts; consider an alternative policy. Supposing the South Africans, the people who would be worst affected by the IMF sales, had professed to be stoically unperturbed by the prospect. Suppose they had broadcast the above arguments and said that they would welcome the sales of IMF gold, because although it might be painful to them in the short term, any diminution in official reserves would hasten the coming boom.

The markets would have leapt to attention. Here were the people who should know, who had most to loose, greeting the IMF sales as a valuable step on the path to a higher gold price. This perverse attitude would have attracted great attention from the financial press, and those invaluable persons, investment advisers. Both would have started to suggest to their readers and clients that they get some gold into their portfolios before the rush. The boom would then have been off to an early start, and without the need for me to go without a decent night's sleep for the fourth day in succession.

Chapter Seventeen

SHEER CONJECTURE

'There is a great deal of evidence, which I have personally studied a good deal myself, to indicate that what I have said is correct; there is no irrefutable evidence of the origins of the ruins at the present time.'

—The Minister of Internal Affairs of Rhodesia, Lance Smith, quoted in that government's *Hansard*, 77, 20, 1969. col. 845.

The ruins of Great Zimbabwe were not the first relics of the past to become a political football; think of the Elgin marbles and the Stone of Scone. But they are the only ones whose importance was such as to have given their name to a country. The Hansard quote above illustrates the fraught atmosphere that surrounded the question of their origins. After all, if Great Zimbabwe had been built by the blacks then it could be used as a national symbol by them; a focus for revolt . . .

Which it was. It did not matter that the Rhodesian Front government wished to cast a decent cover of uncertainty over the matter; as far as the nationalists were concerned it had been built by their forebears. Which it had been. There is no serious argument today over this any more; the ridiculous part is that it took so long for the truth to be accepted.

• • • •

Lord Randolph Churchill, Winston's father, had thought to become rich from gold mining in Mashonaland, but was swiftly disillusioned soon after arrival. So he made a rather grumpy tour of the new territory, concluding that 'the question presents itself, and is found to be almost unanswerable, What is to be done with this country?'[103]

His views were affected by two things. Firstly, his expectations

Sheer conjecture

had been inflated when he went through Johannesburg on his way northwards. Here he went down the Robinson mine, whose ore at that time ran at two ounces (62 grams) to the tonne. J. B. Robinson, you may remember, was first off the block when the Witwatersrand proper was stumbled upon at Langlaagte farm, and this was his reward. The mine was fabulously rich; in Zimbabwe the average gold mine grade for most of the twentieth century was probably about a miserable tenth of that. Secondly, Lord Randolph was already labouring under the syphilis that eventually killed him: he died just three years after the publication of the book from which this quote was taken, aged only forty-six.

But to the point of this chapter, the most interesting statement in the book come from Lord Randolph's mining consultant, an American, Henry Cleveland Perkins. The latter commented on the gold mining of the 'ancients' as follows:

> These old workings are of a very singular and persistent character throughout the district, consisting for the most part of circular shafts varying in depth from twenty to eighty feet and not more than thirty to thirty-six inches in diameter, which have been sunk at all sorts of distances apart, in many cases not more than one foot, and in others as much as fifty or a hundred.

This describes exactly the present-day artisanal workings in the remoter parts of West Africa. Miners with easy access to picks and shovels, such as many in Ghana and Tanzania, generally sink square shafts sometimes using, if they can get them, 'jumper bars'—old drill rods that they hammer downwards to break out the rocks. Those without such tools, or those still enmeshed in traditional ways, such as the *orpailleurs* in parts of Mali, use only small locally-made iron chisels whose blade is periodically sharpened and re-hardened at the surface. With this they cut a narrow circular shaft by squatting at the base and driving the chisel round and round in the floor, using a hammer which is itself sometimes of stone.

These artisanal miners prospect by searching for quartz outcrops, then crushing and panning samples. A surprising number of 'wildcat' pits are dug by prospectors who have a hunch. Some indeed seem to have a knack for finding rich reefs; these 'minefinders' become celebrated amongst the 'sokomotos', 'galamsey' and *orpailleurs*, and their movements are closely followed. Once gold is found, a rush occurs and the strike of the reef is quickly delineated by a line of many pits such as Henry Perkins saw, sited close together. Otherwise there are only the odd 'exploration shafts' scattered through the area. The miner's objective is to get deep enough to intercept what is hopefully a richer zone where the water table has persisted—the 'supergene enrichment'. So these narrow shafts can go very deep—Mr Perkin's 80 feet is the height of an eight-storey building.

He thought these holes were perhaps fifty to a hundred years old. They were in fact much older, the average age probably being about four hundred years, excavated during the flowering of the country as a major gold exporter. Which brings us back to Great Zimbabwe. This was, unequivocally, a gold-trading centre; that is why its ruins are such an exception to the rule that sub-Saharan Africa has little conventional archaeology. For a few hundred years the people of the country had the opportunity to raise themselves above subsistence level by mining and trading gold. No longer dependent on a semi-nomadic existence of pastoralism and slash-and-burn cultivation, the stasis that occurred with the emergence of gold-trading centres allowed the inhabitants and their rulers to plan ahead for the first time. They began to build.

Ready to hand was a useful raw material. On the highveld where the gold is found, the heat of summer and the frosts of winter cause the granites there to peel off layers of stone: exfoliation is the technical term. These layers are anything from a thin scale to thirty centimetres or more thick, but are usually about ten centimetres, and can be cracked with iron chisels into brick-like shapes.

Great Zimbabwe is only the largest of an estimated 150 ruins in the same style that are dotted over the granite backbone of Zimbabwe. It is likely that there were more; one estimate has it that as many as fifty more were destroyed[104]—pulled apart for their stones—since the coming of the white man. A further 65

Sheer conjecture

stone ruins of a later era and with a slightly different style (Khami) have also been found.

There are antecedents. Present-day Ghana is named after an ancient gold-trading kingdom that thrived somewhere beyond Senegal, probably near Kenieba, where the Malian gentleman who was photographed on his throne of gold bars was operating a thousand years later. In the tenth century descriptions of its great buildings and an elaborate culture reached Spain, located at a centre that has been tentatively identified as Koumbi Saleh in the east of present-day Mauritania. Then a few hundred years later came the Kingdom of Mali, centred in the headwaters of the Niger in Guinea, still an important artisanal gold-mining area. Here, as the easy-to-win gold made men wealthy, a rich and sophisticated empire developed, and dwellings became large and permanent. Finally, the Asante of today's Ghana developed the goldfields there in the eighteenth century, creating an empire with an elaborate capital at Kumasi.[105]

The point is that once a semi-nomadic peoples have reason to abandon shifting cultivation and pastoralism, they are capable of creating permanent structures that are remarkable by contrast with their earlier exiguous dwellings. It is not uncommon elsewhere in the world; once Genghis Khan's descendants settled down they abandoned their tents to create the cities of Karakorum, Shangu (Xanadu) and Daidu (now Beijing).

But, given the history of Southern Africa, it was inevitable that the discovery of these remarkable structures led to a racially polarized debate. If, as all right-thinking (white) men agreed, the blacks could not have built them, then who did? It was a outlook of some antiquity in itself; the Portuguese who first heard of Great Zimbabwe in the sixteenth century, shortly after its demise, and who encountered many other '*mazimbabwe*'* when they ascended to the highveld from the Zambezi Valley, far to the north, were convinced they not been built by the local people.

Indeed, João dos Santos, the priest who has been quoted before

---

* The Shona phrase *dzimba dza mabwe* means houses of stone and this has been corrupted and abbreviated to *zimbabwe*. So, to distinguish it from the rest, we have Great Zimbabwe.

and who worked from 1586 to 1595 as a missionary in the northern part of Zimbabwe (by then the Mwenemutapa's country), recorded that the locals had a tradition—and it is a tale that has been swallowed ever since—that King Solomon's gold came from up there. Alternatively, it was the source of the Queen of Sheba's gold. Or perhaps it was the work of the Phoenicians. In any event the story was too good to die, and was taken up by geographers all over Europe. The rumoured position of the fabled city of Zimbabwe appeared on some early maps.

Still, in the steady glow of the late Victorian era, when progress seemed to be pre-ordained, it certainly came as a shock for the first European hunters and explorers (one and the same usually) to encounter this stone-built 'lost city' amongst the scattered and primitive inhabitants of the highveld. That mankind can go backwards as well as forward seemed to have been forgotten. Also forgotten, lying unread in the archives of Lisbon, were comprehensive chronicles telling of the Portuguese efforts to settle Zimbabwe in the sixteenth and seventeenth centuries, and what they found there. What had not been forgotten was the potent myth of King Solomon's mines.

Contrary to a general impression, there is no evidence that the Portuguese actually saw Great Zimbabwe. The first recorded European to glimpse the overgrown remains was a German-American trader and hunter, Adam Render, in 1868. However, the first details came from someone we have met before, the redoubtable but excitable German geologist, Carl Mauch, whom we last met crushing and panning one of Lieutenant Lys' pudding stones on the Witwatersrand-to-be in Chapter 7. He had been fired up by information on the ruins accumulated by the Reverend Alexander Merensky, a missionary in what was then the Transvaal. Merensky's son, Hans, was to be another famous geologist; he found the diamond fields at the mouth of the Orange River and the deposits of phosphate-rich rock at Palaborwa that have given South Africa a supply of fertilizer that should be good for a thousand years.

Mauch made his difficult way northwards and stayed for nine months with Adam Render, although he was only able to make

Sheer Conjecture

three visits to the ruins in this time. It appears that the local Karanga—a Shona clan—had their suspicions of him. But what he found and heard was enough to get the whole Solomon-Sheba-Phoenician thing going like wildfire again. A piece of wood from the ruins that resembled that of his pencil set him off in almost incoherent excitement. In his diary he identified the timber as—

> ... cedar-wood and from this it cannot come from anywhere else but the Libanon. Furthermore, only the Phoenicians could have brought it here; further Salomo (Solomon) used a lot of cedar-wood for the building of the temple and his palaces: further: including here the visit of the Queen of Seba [sic] and, considering Zimbabwe or Zimbaoe or Simbaoe written in Arabic, (of Hebrew I understand nothing), one gets as a result that the great woman who built the rondeau could have been none other than the Queen of Seba.

According to Mauch the local Karanga, themselves fairly recent arrivals to the ruins, thought that the building had been done by whites long ago. Handily, this gave the occasion of Cecil Rhodes' first visit a chance to introduce him to the local chiefs as 'the Great Master' who had come 'to see the ancient temple which once upon a time belonged to white men.' It was an nice irony that the 'minor Karanga chief' controlling the area of Great Zimbabwe, whom Mauch encountered in 1871, happened to be called Mugabe.

The little museum at Great Zimbabwe suggests that the Karanga themselves were the builders. It is certainly possible that their forebears were involved, but they would not have called themselves Karanga. Once the accessible gold had been exhausted and the traders had left, the local people would have had to revert to a drifting existence of pastoralism and shifting cultivation, and the Great Zimbabwe area would have been abandoned. The ensuing migrations and invasions must have scattered the builders far and wide.

There are two main areas of ruins at Great Zimbabwe. One is the Central Enclosure (in my childhood it was called The Temple). Interestingly—shades of Rider Haggard—Carl Mauch found that it was known to the local Karanga people as 'the house of the great woman'.[106] It is a multi-walled enclosure, roughly elliptical, 50 to 150 metres across, the outer wall alone being estimated to contain some 10,000 tonnes of granite blocks. Inside are an incoherent jumble of secondary walls (some of them substantial), a separate smaller enclosure and the famous conical tower. The latter design is repeated *ad nauseam* at various hotels and publicity centres in the area and, bizarrely, the air traffic control tower at Harare's new airport.

The second group of structures is on a narrow ridge of massive granite boulders, about 100 metres high, overlooking the Central Enclosure. Between the boulders, stone walls and enclosures have been built. It has the look of a defensive position, and certainly that was what Chief Mugabe and his people were using it for; they lived in fear of Matabele raids. It is now called, boringly, the Hill Ruin, but we learnt to call it the Acropolis.

Unfortunately, once serious archaeological work was begun the story about its foreign origins proved difficult to sustain. Even the earliest investigator, J. Theodore Bent, an amateur who was recruited by Rhodes' BSA Company in 1891, found little that would enable him to establish a non-African provenance beyond some bits and pieces of glass and ceramics dating from the late Middle Ages. Nothing daunted, he assumed the soapstone carvings and other features of the ruins indicated that they arose from a 'northern race coming from Arabia.'

A fair amount of looting occurred thereafter, both at Great Zimbabwe and in the other *mazimbabwe* scattered around the country. Almost no gold had been found by Bent, but the other sites yielded considerable ornamental work, mostly, if not all, on skeletons in burial grounds. Items of gold do not normally go long undiscovered at places where there are no taboos against taking them.

It is hard to know how much was actually found. The average amount of gold reported from each ruin was less than ten ounces, but the BSA Company had a right to 20% of the finds. Looking at the gold smuggling that went on during the early years of the country on account of high mineral royalties, it seems likely also that the total from grave-robbing was similarly under-reported.

Sheer conjecture

In this light, Great Zimbabwe was almost certainly a disappointment; it was not built as a mausoleum, it had a much more mundane function. Major Sir John Willoughby, a wealthy aristocratic type, excavated enthusiastically in some of the ruins without turning up more than the same sort of pottery shards that Bent had. After him came Richard Hall, a journalist and PR man. He was supposed to be a curator, but he dug up great areas of the ruins to remove 'the filth and decadence of Kaffir occupation'. Eventually he was sacked, but not before he had managed to obliterate much of the stratigraphic record.

The first professional archaeology undertaken at Great Zimbabwe was in 1905, ironically under the auspices of Rhodes' trustees (he had died in 1902). The archaeologist was David Randall-MacIver, who was chosen by the British Association to undertake the work. He concluded that the pottery shards he had found at the lowest strata were of the same pattern as the present-day Karanga used. This was inflammatory stuff and Richard Hall rebutted these findings with a book—*Prehistoric Rhodesia*—where he wielded the racial cudgel vigorously; ' ... the decadence of the native ... is admitted by all authorities.' It was a bestseller in Rhodesia.

The next investigation was an all-woman affair in 1929, headed by Gertrude Caton-Thompson. She again could find no evidence of an exotic origin for the buildings; indeed the very absence of all but a few pieces of Chinese porcelain and some Indian beads made the dating of the ruins very difficult. But unequivocally, it was of African origin and less than 1,000 years old.

So that was that, then. The —we —whites rationalized the thing away; after all she was new to the country and one of those mannish sort of women who think they know everything.* Nothing much else was done for another twenty years.

Then, in the 1950s, the whole question blew up again.

---

* Rhodesians' dark suspicions seemed to be confirmed when Gertrude Caton-Thompson joined the adventurous Freya Stark on a quarrelsome expedition to the Hadhramaut in 1938. Sadly for scandal, there is no evidence that these ladies ever felt anything more than dislike for each other.

All poor together

Technology had moved ahead, and the Carbon-14 dating—a very recent development at that time—of wood taken from a drain through a wall indicated that it was from a tree felled in about AD 650. So there, the wretched Caton woman had been wrong after all; the date was before any major movement of Bantu peoples into Southern Africa.

However, the archaeologists (by now not all of them female, and not all of them from outside Rhodesia) stuck to their guns. The walls had unequivocally been built after the tenth century, but the wood (which was from a long-lived species, *Spirotachys africana*, known as tamboetie, or African sandalwood) was heartwood and so may have effectively become moribund long before the tree itself was cut down.

Frankly this was unlikely. The wood is certainly durable, as it is one of the few that resists the white ant. But for it to last unaltered for 300 rainy seasons before being incorporated into the wall? And so the Phoenicians, King Solomon, the Queen of Sheba, all the usual suspects, were rounded up once again. Their resurrection was celebrated in 1973 by another book full of vigorous arguments, *The Origin of the Zimbabwean Civilization*, by Richard Gayre.[107] He was a Scot—his full title was Richard Gayre of Gayre and Nigg—and took a certain sardonic pleasure in skewering the experts—

> In order to escape from the conclusive Carbon-14 evidence, those amongst the later archaeologists who have constituted themselves the pro-Bantu school have been forced to ludicrous shifts to explain the evidence away ... Apart from knowing nothing of the needs of builders, for whom old, as distinct from seasoned wood, is of no use, and the fact that the forces of nature would have destroyed it by means of fire and insects, the whole of this special pleading is verging on dishonesty ...

But then the pendulum swung once more. Thomas Huffman of the Department of Archaeology at the University of the Witwatersrand (oh, what don't we in Africa owe to gold?) worked at Great Zimbabwe in 1968 and 1970, at a time when the Carbon-

Sheer conjecture

14 dating technique was being refined to much greater levels of accuracy.[108] A re-examination of the tamboetie wood now showed that it dated from about 1340, spot on as far as the stratigraphic date indications were concerned, and long, long after the Phoenicians and the rest. I cannot recall hearing this in Rhodesia at the time; it was presumably not exactly welcome news to the government of the day and the local press was under an effective censorship.

Richard Gayre and the other proponents of the school of thought that said that foreigners, Phoenicians, Arabs, or what-have-you, must have built Great Zimbabwe and worked the mines, point out that the Shona people were not interested in mining. This was true; the modern mines of Rhodesia were worked very largely by immigrants from Malawi and Mozambique (although the latter were often of Shona stock themselves). Since independence, however, the situation has changed; many of the older miners have retired or died and their places have been taken by local people.

A lack of a mining tradition does not mean therefore that people won't mine if necessary. In Kenya the Masai *moran* who were so confrontational when dealing with us as prospectors on their land, are also working underground in the mine that was developed there. Their sinewy physique turned out to have both the strength and the aptitude for hard physical work in demanding conditions. Something similar probably happened with the forebears of the Shona in the mines of medieval Zimbabwe.

So today the archaeological picture is of a sequence of inhabitants of Southern Africa. First there were the hunter-gathers, of which the San (the bushmen) and the Hottentots, collectively known as the Khoisan, are the lingering survivors. Then into the area in about the fourth century AD moved the first Bantu people—the regional distribution of the pottery they left behind suggests a common origin in Katanga—who were farmers, and had mastered iron smelting. They seemed to have no cattle, only sheep or goats, and they limited their agriculture to the light sandy soils of the granite areas. Their relics—pottery, ironwork—have been found in the lowest levels of the Great Zimbabwe area,

All poor together

predating the wall construction there by many hundreds of years.
After that, in about the twelfth century AD, arrived a further wave of iron-age Bantu peoples who were cattle-owners, more pastoralists than farmers. However, while goats and sheep can live off the grasses of the sandy leached soils of the granites, they tend to be poor fodder for cattle. It may be guessed that the newcomers colonized the areas of richer vegetation associated with the heavier red soils of the greenstone belts. Here most of Zimbabwe's gold is found as well as today's best farming areas.

What follows is pure hypothesis, but it seems probable that only when this occurred did the true gold potential of the land become apparent. Now settled on the best spots for cattle, the effort had to be made to till the soil for millet and sorghum, the indigenous African crops. Sometimes the hoes would turn up small nuggets—there have been numerous recent discoveries made in just this way, such as Nyambegena in Tanzania and Karimtanga in Burkina Faso—and even a naïf with no knowledge of gold would guess that here was a metal of value.

A modest gold trade with the coast probably predated this realization. The granites of the areas colonized by the earlier invaders do have gold reefs in them, but usually these are small and low grade. In addition these rocks do not weather as do the greenstones, making the extraction of the metal doubly difficult.

By contrast the greenstone belts are rich, well weathered and numerous. There are something like thirty-one of these in the region, of which twenty-two alone are in present-day Zimbabwe—a precise count is irrelevant due to their intermingling and irregularity—and while in general they are elongated, they are blobs rather than belts.

But for local miners the gold in greenstone was not inexhaustible either. Just as with most present-day artisanal mines in Africa, deposits could be found, exploited and abandoned in the space of a year or so. Their life was generally brief because production stopped with the depletion of the supergene enrichment zone or the reaching of the water table or the attainment of hard unweathered rock (and often all three at the same time). The transitory nature of artisanal mining and the extensive dispersion of the greenstone

Sheer conjecture

belts across the region meant that Great Zimbabwe was only one of a sequence of gold trading centres during Zimbabwe's first mining era.

The earliest was Great Zimbabwe's predecessor. It was in the south, on the southern bank of the Limpopo at a place called Mapungubwe (a literal present-day Shona translation might be 'tamed stones'). This evidently started as an ivory centre, but then it became a place where gold could be traded—it was close to the gold belt of present-day Botswana—and the richer sort of inhabitants began to live behind stone walls. However, Mapungubwe was abandoned in the latter part of the thirteenth century, and the focus shifted northwards to Great Zimbabwe.

The interesting question to a miner who has had much to do with present-day artisanal gold mining in Africa is not so much who built Great Zimbabwe as why was it built where it is? For the ruins are not on a greenstone 'gold belt'; the nearest is about ten kilometres away, while the nearest gold mines are a further five kilometres beyond that.

The answer to the siting question is actually simple. The Zimbabwe Ruins sit at the entrance to the highveld, colder but healthier than the lowveld, near the headwaters of a usually perennial river—the Mutirikwe. The Swahili-speaking traders—Muslim Africans themselves; the Portuguese, you may recollect, called them Moors—could get there from Sofala on the coast after a march of perhaps twenty days. This with no danger of thirst, and a certainty of game for the pot *en route*. The traders would follow the south-westerly trending line of the Busi River from the coast near Sofala up to where it petered out at the southern limits of the mountains that separate Mozambique from Zimbabwe. Then over the hills and down to the Save River, just on the other side. (In the Nguni-based language of the nineteenth-century Matabele invaders, which has b in place of v, this was called the Sabi, and inevitably it was argued that the name was derived from Sheba.) Then from the Save they travelled up the Runde River (at the junction of the Save and the Runde is another set of stone ruins) and then up the latter's Mutirikwe tributary which would take the traders to within ten kilometres of Great Zimbabwe. Hence the

All poor together

date palms by the Mutirikwe River next to Robert Kennedy's Renco mine (chapter two).

For traders this was probably the traditional route for a long time, and Great Zimbabwe arose as a response—to intercept them. Almost certainly once it, or more likely some simpler predecessor, was in place they were not allowed to penetrate farther. The country must have been essentially a unitary state by the end of the thirteenth century, if the consistency of the walling of over a hundred *mazimbabwe* of the era is anything to go by. The ruler of such a country could ensure—or try to ensure—that the gold produced was funnelled through his stone city to capture the benefits of its sale, just as the Reserve Bank of Zimbabwe today demands—or tries to demand—that all the gold won in the country passes though its hands.

The number of mines was very great. One estimate is that something like 4,000 mine sites were developed between the eleventh and the fifteenth centuries, with an output of over 600 tonnes, perhaps a third of the world's total production in that period.[109] These numbers are not surprising; when Tanzania liberalized its gold trade laws in 1991, so destroying the black market, it turned out that artisanal production, using techniques not greatly improved from those of medieval Zimbabwe, was about four tonnes a year from about three hundred sites. At the time this gold was worth about U.S.$50 million. Gold, it was discovered, was one of Tanzania's biggest foreign exchange earners, and that from a country with, Buckreef apart, no formal gold mines.

Hence if, as the archaeology indicates, Great Zimbabwe was a gold trading centre for perhaps 250 years, it would certainly have been possible for several hundred tonnes to be mined and exported through this route.

As structures go, the dry stone walling technique using blocks of exfoliated granite is not a particulary stable one. Numerous collapses have occurred since the modern rediscovery of the *mazimbabwe*. Great Zimbabwe was declared a World Heritage Site some years ago, and a British-funded project to stop the walls from falling down has been on the go since the late 1980s.

Sheer conjecture

However, the demands of civil engineering and those of authenticity clash. It has been laid down as a general principle of architectural conservation that modern structural engineering solutions cannot be used. So concrete grouting is out, and the stones have to be replaced just as they were before. This is perhaps tendentious; much of the Great Enclosure has already been rebuilt, in the period 1911–1948, and as a consequence some of the most admired parts, including the entrances to the Great Enclosure, are recent restorations.

While the building system used led to the *mazimbabwe* being sited mainly on granite and gneiss (strictly orthogneiss, a metamorphosed granite), they have been found to be well placed in relation to the mines on the adjoining greenstone belts, and even more closely linked to sites where gold was smelted and fabricated.[110] This is one of the main arguments for gold being the spur behind the creation of Great Zimbabwe and its offshoots. This position has been challenged by two main schools of thought, although neither of these dismiss the importance of gold.

One says that the cattle-rearing newcomers who built the *mazimbabwe* did so because they were getting rich not just through gold but also by using a transhumance technique to fatten up their cattle. Transhumance is the seasonal migration of livestock, and the argument goes that during the dry season the herders moved their cattle to the lowveld to live off the nutritious leaves of the mopane trees there, and then, when the rains came, were herded back up to the highveld to take advantage of the fresh grass cover.[111]

As a hypothesis it has some merits. The dry season is the only healthy period in the lowveld; during the rains malaria and sleeping sickness from the tsetse fly are endemic there. But the distances are great, often fifty kilometres or more between the two ecologies—there is usually an extensive middleveld in between—and the greenstone belts provide good grazing throughout the year on the highveld anyway. In addition, peoples who are semi-nomadic in this way do not normally bother to build permanent structures.

The other theory has more weight. This is that iron—particularly

All poor together

in the form of iron hoes—was almost as precious as gold to the builders of Great Zimbabwe.[112] The site is about 80 kilometres from a mountain of iron ore known as Buchwa, that was used until recently as a feedstock for the country's steelworks. *Mazimbabwes* have been found at Buchwa, and there is evidence of iron smelting having been carried out in the Central Enclosure as well as at numerous other stone walled sites.

There are also indications that copper and bronze were smelted at Great Zimbabwe. Copper was mined and smelted widely in prehistoric Zimbabwe, as well as on what were to be the Copperbelts of Zambia and the Congo. A casting mould of the distinctive 'x' shape that was used for this metal was found in the ruins. Malachite, a copper oxide whose dramatic green colour makes it easy to spot and which can be smelted readily with charcoal, has been found on the 'Zimbabwe Claims' about 20 kilometres west of Great Zimbabwe. For bronze (an alloy of copper and tin) the 'Bikita Tinfields' lie about 70 kilometres in the opposite direction.

There is therefore probably some validity in the 'precious iron' idea, particulary as the bulk of the population was, one way or the other, grazing cattle or mining gold on the heavy soils of the greenstone belts. To mine and cultivate there the inhabitants would have needed iron implements. Nonetheless, iron goods outside Zimbabwe were common, and it could hardly have been an export item. Indeed, some iron items found at Great Zimbabwe appear too sophisticated to have been made locally. There was only one metal that would have attracted the traders all the way from the coast.

(Well, that may not be quite true. In the early 1930s the great anatomist Professor Raymond Dart was working with an Italian scientific expedition in the area of what is now Kabwe, in Zambia. Here he found the remains of an archaic *Homo sapiens,* Broken Hill man, a modest milestone in the route to the elucidation of man's ancestors. Professor Dart also recorded that they discovered an ancient manganese mine—so ancient that it had been mined with quartz hammers. Now this sounds ridiculous. Manganese is a metal used for hardening steel, an attribute not discovered until the nineteenth century. So what was Neolithic man doing with it?

Sheer conjecture

The answer has to be that it was used as a pigment and a pottery glaze. The black manganese oxide, pyrolusite, when ground fine, is ideal for this purpose. But why was so much needed? From the excavations it appeared that several hundred tonnes had been mined, indicating a trading activity far beyond the needs of a few scattered hunter-gatherers. Professor Dart suggested that it had been exported to ancient civilizations, to Egypt. Sumer and Babylon, all of whom had mastered the use of manganese glazes on their pots.)

To return to Great Zimbabwe, eventually here too, just as with Mapungubwe, the limits were reached of the gold that could be artisanally mined in its hinterland. In the sixteenth century the visits of the traders became scarcer, and they moved their routes southward again, to come up the tributaries of the Limpopo to the gold belts of what is now Matabeleland. A fresh burst of walled settlements erupted in this area, of which the Khami ruins (near Bulawayo) are the representative type. They differ in detail from their predecessors, and it is thought that they arose from the activities of an unrelated peoples—the Torwa.

North of Great Zimbabwe, another new kingdom, that of the leader known as the Mwenemutapa (perhaps from Mwana Motapa, child of god), arose. This was centred close to where the central plateau fell away into the Zambezi Valley. In these northern gold fields the Great Zimbabwe-type walled centres were few and far between and, unlike the passive, anonymous rulers of Great Zimbabwe, the Mwenemutapa led an aggressive, expansionist state, whose reputation and influence eventually reached to the coast.

Here were new traders, carrying firearms. They were the Portuguese, who under Vasco da Gama had rounded the Cape in 1498 and built a fortress at Sofala to capture the gold trade. Unhappily for them their arrival probably coincided with the falling-off in the output coming down the Busi from Great Zimbabwe. Foiled, they sought to find a river route into the interior, and discovered the Zambezi. Establishing a base at Angoche, near its mouth, they brought to the valley of the great river the same mixture of violence, religion, slavery and

All poor together

miscegenation that the Spaniards introduced to South America. There was one critical difference. In America the diseases of the old world slaughtered off the majority of the inhabitants of the new world, leaving the way open for foreign domination. In Africa it was the other way around. Here the diseases of mankind had lurked for centuries, in varieties unfamiliar to the immune systems of the invaders. The Portuguese thrust was dissipated by malaria and sleeping sickness. The survivors were mainly their African slaves, whose armed and barely-controlled bands (*chikunda*) caused havoc amongst the Mwenemutapa peoples.

But now Zimbabwe had entered written history, and the fourth phase of its gold-trading began. This centred not on one but on half a dozen *feiras* (fairs, trading posts) the Portuguese created, at gold-mining areas stretching in a rough arc across the north-east of the country—Mudzi, Bindura, Mount Darwin, Karoi. Here too, stone buildings were created, but there were important differences; they were rectilinear and they had ditches and earthworks around them. They were on the greenstone belts, not the granites, and so their construction was from the irregular metamorphic rocks of that geology, roughly trimmed. They had loopholes for firearms. There is little argument about whose influence caused these structures to be built.

The Munhumutapa state, much weakened, was supplanted by one of its vassals, the *Changamire,* who adopted the ruthless methods of the *chikunda* to drive the Portuguese away from the highveld and into the heat and fevers of the Zambezi Valley. The Portuguese themselves did not last in this environment; they gave the surviving *mestijo* leaders of the *chikundas* land grants (*prazas*) in return for keeping a nominal Portuguese occupation, which they did for another three hundred years. In the meantime the victorious *Changamire* headed southwards, supplanted the Torwa and set up the kingdom of the Rozwi (the destroyers) which extended beyond the Limpopo. This lasted, its hold slowly weakening, until the nineteenthth century. Carl Mauch found a Rozwi at Great Zimbabwe who described ceremonies performed there by his father. But by that time the Matabele had broken away from the Zulu kingdom and come storming northwards, finally

Sheer conjecture

dispelling the Rozwi's residual authority. By then all the gold had gone as well.

• • • •

So that was how the saga of Great Zimbabwe ended. But how, in the light of this knowledge, did it start?

The sequence may have been this. As soon as gold began to trickle out of what is now Zimbabwe, the Mutirikwe River became the best route in for the 'Moorish' traders. The granite expanses around Great Zimbabwe marked the end of the ascent to the highveld and made a natural halting point. It is cold and windy there, particularly in winter, and for the traders accustomed to the steamy heat of the coast, a miserable place to wait.

Even before Great Zimbabwe was built they would have probably waited there. A feature of artisanal gold, and even more of gemstones, is the number of transactions that occur. The buyer at the mine site—and there may be as many buyers as miners—is typically an agent for a trader, who in turn will sell on the gold to other intermediaries in the chain before it arrives at its final destination—nowadays often a refinery in Europe.

So the traders from the coast would not have expected to meet miners. They would be dealing with local gold trading magnates, who were at the centre of a network of suppliers. Transactions would be made at a place convenient to both, and as the trickle became a flood, that was where Great Zimbabwe eventually developed. To keep warm in this bleak environment, and perhaps to discourage thieves, they may have supplemented the temporary timber and mud huts they built with stone walls. So it could have begun, as a genteel competition as to who could build the most impressive accommodation. As the trade grew, the king moved to this, the centre of economic activity in his domain, and what house-agents today would call a prestigious development was indicated.

The hunter and early wild-life enthusiast, F. C. Selous, who led Rhodes' pioneers into Zimbabwe, had no doubts. He consistently said that he saw no reason why the local people could not have created Great Zimbabwe.[113] Finally, one important argument. Whoever built it—and whoever dug the mines—could not have been lettered, otherwise our information would not have come solely from the uncertain, second- and third-hand, accounts of outsiders.

To state it bluntly, Great Zimbabwe must have been erected by Africans. It has nothing rectilinear about it, no straight lines. It is inchoate; the Western eye and brain searches fruitlessly for any suggestion of order in its layout. It was created, not designed, one man waving a hand in this and that direction, later another elaborating on it just as capriciously, and so on. Over perhaps fifty years or more the builders improved their scale and technique of bricklaying, culminating in the massive north-east section of the outer wall of the central enclosure, with its chevron frieze. When the gold ran out and its usefulness was over it was abandoned and left to fall apart.

What about the stories that João dos Santos recorded, that the local people had a tradition that the mines he saw in the north of the country were the source of gold for Solomon and Sheba? It was an idea absorbed from the 'Moorish' traders they had been dealing with up to then, the Swahili speakers who came up from the coast. Islam, after all, shares with the Old Testament much of the Middle East's history, including tales of the wealth of Solomon and Sheba.

There is no great mystery; that lies elsewhere.

* * * *

On an evening more than forty years ago Jumbo Harloe and I were stumbling down a wet hillside in the foothills of the mountains of Nyanga, whose granite heights form the northern wall of Zimbabwe's eastern border with Mozambique. The sky was lowering with more rain to come, but we were already soaked, both from the waist-high grass and from three nights spent in a tent whose canvas was not up to the torrential storms that had pounded it. We were trying—and failing—to walk in a straight line over high empty country from the peak of Nyangani, the highest mountain, to the village that is itself called Nyanga (except that in those days all these names were prefixed with an Nguni-dialect 'I'—Inyanga, Inyangani).

I don't know about Jumbo, who was a stout, cheerful type, admirable as a companion in this sort of fix, but I was pretty miserable. It was not just the drowned-rat feeling. We were here because I was trying to get my Scout's hiker badge, this was the second attempt, and we—I—was going to fail to get to the objective. Without speaking much about it, we had agreed an hour or so ago that when we reached the road, we would turn left, to hitch lifts back to Salisbury and home, rather than straight on over and into the further wet night needed to complete the march. From this distance in time it was a silly thing to set out to do, just pitching ourselves as a feat of endurance against the rugged high grasslands in the middle of the rains.

Staggering downwards, we suddenly came to an abrupt fall, where stones had been piled up to hold back the soil. Despite our sodden packs it was an easy scramble down—about a metre—but after a couple of paces there was another drop in front of us, and then another and another. We found ourselves making our way down a hillside whose face had been cut up into narrow terraces, some as little as only a metre wide. After a score or so of these we reached the base, waded thigh-deep through the swollen stream at the bottom and then ascended the far bank. Here again there were more terraces, wider than the first due to the gentler slope, but difficult to mount quickly with our loads and in our exhausted condition. We stopped and looked around. Their regular lines extended on either side as far as we could see, precisely contoured, covered with rank grasses and empty of life. In front they covered our route up the hill. We were not going to make the road that day.

Great Zimbabwe may be formless but there are ruins in the country that are the result of a plan; they have precise straight lines, they have, or had, a function and much more effort went into them. In those high foothills of the bleak eastern border country lies a vast—there is no other word for it—system of stone-built terraces. It covers 6,000 square kilometres and with them are associated complex irrigation channels and many, many stone ruins, one group of which—the van Niekerk ruins—extends for over fifty square kilometress. Here was a static society with a vengeance.

These ruins have layouts whose sections are almost as opaque in function as Great Zimbabwe. They are not as well made—exfoliation of the Nyanga granites and gneisses is not as prevalent—and they probably predate the roughly rectilinear forts (some with either low loopholes, or maybe high drain holes) that sit atop strategic hills in the area. It might be guessed that the *chikunda* made these forts (although there was no gold for a hundred kilometres around). But these other ruins aside, who made the terraces, and when? What was grown there? And why did they leave?

There are some tentative answers currently in place. It is now thought that probably the ancestors of the local Unyama people were responsible.[114] They were surrounded by hostile groups and had no access to flatter territories, and so had to make do with the steep, cold and empty hillsides they had been confined to. Carbon dating indicates that construction was under way in the sixteenth century.[115] Like Great Zimbabwe, like the empires and kingdoms of Mali, Ghana and the Asante, once an African people had reason to remain in one spot, they started to build.

But this still leaves major questions. What crops were grown in this un-African environment? The only other places on the continent where such extensive terracing is found are in Burundi and Ethiopia, on rich volcanic soils in generally hot climates. There, respectively, rice and teff (the Ethiopian staple cereal) are farmed. But on the terraces of Nyanga, the soils are of granite origin and generally poor, the climate cool to cold. Was that immense effort made just to crop for a few seasons while building a new set of terraces farther on? And why such sophisticated irrigation works? What, in this wet corner of Zimbabwe, required so much water? Nowadays only winter wheat gets this treatment. Was it that, uniquely, they were justifying the effort required to create terraces by growing two crops of maize a year? And finally, why, at some time in the eighteenth century, did the builders stop and abandon the hillsides to their mists and winds?

Sheer conjecture

A couple of months after we got back to Salisbury, Jumbo went down with polio, and the last time I saw him, many years ago, he was still limping. He insisted it didn't have anything at all to do with those cold, wet days struggling up and down the dripping hillsides and terraces, but as I said, he was an admirable sort of person.

Chapter Eighteen

SOMETHING IN THE AIR

Moreover, as the miners dig almost exclusively in mountains otherwise unproductive, and in valleys invested in gloom, they do either slight damage to the fields or none at all.
— Georgius Agricola, *De Re Metallica*, 1556.

Agricola's dignified defence in the face of attacks by the environmental lobby of medieval Thuringia was echoed in more detail 440 years later by Robert Wilson, Chairman of Rio Tinto Plc. He pointed out that his company's global output value of U.S.$7.7 billion in 1997 involved the disturbance of 45,000 ha of land (most of which would be restored to its previous condition). To create the equivalent turnover from hi-tech farming would require at least 13 million hectares, nearly 300 times as much area.[116] Yet, while Rio Tinto's annual general meetings often resemble a riot due to the noisy interventions of environmental activists, very few people are agitated about the loss of biodiversity, the contamination of water and the fuel and electricity requirements that farming incurs.

Mr Wilson had special reason to be caustic about environmentalists. A whole NGO, calling itself Partizans ('People Against RTZ and its Subsidiaries') is dedicated to attacking his company on the environmental front. In 1991 they published a book called simply 'Plunder'.* The foreword was provided by Al Gedicks, at the time director of the 'Centre for Alternative Mining Development Policy' and professor of sociology at the University of Wisconsin. He wrote:

> Mining, by its very nature, is an assault on the physical, social and cultural environment. When this assault is

---

* Copies are obtainable, price £6.00, from Partizans, 218 Liverpool Road, London, N1 1LE, Telephone: 0171-6090-1852.

organized by one of the most powerful, arrogant and racist corporations in the international mining industry, the results can be devastating.

Not many farmers get to experience that sort of denunciation. Mining companies have come to make a distinction between 'light green' and 'dark green' environmentalists. The first group accepts that economic growth cannot avoid industrialization and the need for regulation of pollutants. The 'dark greens' believe that a way of life founded on consumer values—which implies industrial and mining activity—causes environmental damage. By that standard, Partizans are very dark indeed.

The irony is that there are abuses worth attacking, but because they are not environmentally 'sexy', they tend not to get much done about them until serious damage starts to appear (perhaps persons of a nervous disposition should read no further).

Dust is one of these. For a century or so, it has been apparent that silica dust—the sort you get from gold mines working quartz veins—causes a chronic lung condition called silicosis. Since about 1970 it has been a given amongst miners and their doctors that silicosis could eventually lead to lung cancer if the sufferer was not removed from working in a dusty environment, particulary if he smoked.

In every country with a significant mining industry (in Africa this means just five countries, South Africa, Botswana, Zimbabwe, Zambia and Ghana), all employees who work in potentially dusty occupations, such as underground and in mills, are required by law to have their lungs x-rayed annually, along with a general chest check-up. These countries have a government-sponsored medical organization usually known as a pneumoconiosis board or centre (after the generic name for diseases from dusts) which does this. Elaborate underground arrangements, including minimum ventilation rates and the use of water during drilling, are mandated by the mining regulations to curb silicosis. In Southern Africa the card that certifies you fit for work in such places is known as a red ticket. Without a red ticket you can't get or stay in a mining job.

Yet the malign nature of silica-rich dusts has been largely ignored by the environmentalists. They had an excuse; it was also ignored in some official circles, perhaps because it is not normally a rich-country disease. Silicosis is a concern of the International Agency for Research on Cancer (IARC), which decides what is safe and what is not in such matters. Yet it was not until October 1996 that it promoted crystalline silica (the type found in mines) from category Class IIB ('a non-malignant pathogen') to Class I ('a known human carcinogen'). Even then agreement was not unanimous: the vote was 10 to 7. This was despite the evidence of a number of studies that confirmed the greatly enhanced risk of lung cancer that sufferers from silicosis faced.[117]

This indifference may have arisen from the apparent rarity of silicosis in the developed world. In the United Kingdom there are thought to be only 30–50 silicosis cases yearly, compared to about 35,000 new lung cancers. This is almost certainly misleading; the overwhelming number of smoking-induced lung diseases and the known vulnerability of smokers to silicosis must mean that there are many hidden cases.[118] In gold-mining countries like Zimbabwe silicosis is unquestionably a major problem; one mine manager on a small and well-run operation confided to me that he had run out of non-dusty jobs to put his affected workers into. As far as he knew all of those affected were smokers.

The new classification of silica dust elevates it to the same category of threat as asbestos. As a consequence there has been a great flurry of 'me too' silica dust regulation in America and Europe since 1997, for, as everybody knows, asbestos is a very bad mineral indeed.

Before going any further, it is necessary to note that there are three diseases of the lung quite specifically associated with asbestos. Asbestosis (more generally, pulmonary fibrosis) is caused by asbestos dust accumulating in the lungs to the point where it gives a shadow on an x-ray image, but it is not usually malignant. Then comes lung cancer, which is, but can often be treated. Finally there is mesothelioma, which is both malignant and untreatable; it is a tumour that develops not in the lungs but in the chest cavity, slowly rendering it incapable of expanding the lungs. It is the dramatic upsurge of this inexorable disease that has led to the strict ban on asbestos in many countries.

For a mesothelioma epidemic is raging in the developed world. On 2 March 1995, at a London press conference organized by the British government's Health and Safety Executive (HSE), Professor Julian Peto of the UK's Institute of Cancer Research announced figures which confirmed its existence.[119] Peto suggested that 1 in 100 of all British men now aged 50 could die from mesothelioma.

The epidemiology of the disease was strange. A firm link had been established between 'blue' asbestos and mesothelioma more than twenty years before the announcement, and during a twenty-year period commencing in 1971 there were 183 mesothelioma deaths amongst workers known to be dealing directly with asbestos. What was shocking was that the figure for mesothelioma deaths amongst people outside the industry for the same time was nearly 11,000.

Because the HSE estimated that for every death due to mesothelioma, there are two from asbestos-induced lung cancer, it is thought that a total of 10,000 asbestos-related deaths could be reached annually in Britain. No wonder the stuff has been banned in the European Union.

The results announced by Professor Peto also flew in the face of a 1985 announcement by an international group of experts convened by the World Health Organization in Hanover (Germany). They concluded that: 'The risk of mesothelioma and lung cancer, attributable to asbestos exposure in the general population, is undetectably low; the risk for asbestosis is practically nil.'[120] They were, it seems, terribly wrong.

Yet perhaps not. The asbestos story is the tale of the uncovering of a terrible disease, but it is also the story of the damage that can be done by violent prejudice.

It was the spinning of asbestos into fire-resistant textiles that first brought the risks to the public eye. Between 1890 and 1895, sixteen out of the seventeen workers of a French asbestos-weaving factory had died. By 1899, eleven men, all aged about thirty, who had spent their (short) working lives in an asbestos-spinning factory in the UK, had also died. To achieve these fatality rates the conditions under which they must have worked do not bear thinking about. The numbers involved and the speed with which they died suggest massive lung congestion; asbestos diseases normally have an protracted latency period (during which time there may be a shadow on the lungs, nothing more) of between 20 to 40 years.

Historically, asbestosis as such was first recorded in 1906, from an autopsy on an asbestos worker done at Charing Cross Hospital in London.[121] However, it was not until 1927 that asbestos was recognized as causing fibrosis of the lungs and the term 'asbestosis' was used for the first time.[122] The next step—the discovery of an association of asbestos with lung cancer—occurred in 1935 when the first recorded case was described.[123]

Then the Second World War intervened and asbestos demand soared. So did its associated diseases. By the 1950s the link between asbestos and lung cancer was solid, backed up by extensive epidemiological data. Records from Canada showed that asbestosis sufferers who smoke heavily are over 50 times more likely to develop lung cancer than non-exposed workers who do not smoke.[124,125] In 1964 a strong association was found between American workers installing insulation and lung cancer.[126] In that same year health warning labels became required by law in the United States on products made with asbestos.

The mesothelioma threat was the last to emerge, but the worst by far. It surfaced in 1960, when a powerful correlation was discovered between asbestos exposure and mesothelioma in people living (but not necessarily working) amongst the blue asbestos mines west of Kimberley, in South Africa.[127]

The deadly legacy of the mines there rages on. The doctor charged with developing South Africa's National Cancer Registry believes that when all the data have been assembled, the Northern Cape will be shown to have the world's highest incidence of mesothelioma. During 1997 teams of volunteer nursing students and medical personnel carried out a survey of 1,000 people in Prieska; 280 exhibited signs of asbestos-related lung changes. Over eighty per cent (232) of these had worked on the mines.[128]

But here is a funny thing; Zimbabwe's biggest mines are asbestos operations, yet the doctors there will tell you that in the cases where the x-ray plate shows the characteristic pneumoconiosis shadow on the lungs of a worker, he has previously mined gold, which in Zimbabwe's geology implies exposure to silica dust. Indeed, since new dust-free mills were built at the asbestos mines in the 1970s the doctors are not sure that they have ever seen a true case of asbestosis, certainly no mesotheliomas. However, they do see a strong correlation between lung disease and smoking amongst the workers from both sorts of mines. Hence the concern at the beginning of this chapter that the incidence of silicosis in the West may be seriously understated.

Yet asbestos has been a threat for many years, while crystalline silica is only now being grudgingly recognized as one. Surely, asbestos is far more lethal than crystalline silica? The answer is, as the reader may have guessed, that there are different types of asbestos, and the mineral that Zimbabwe recovers—chrysotile, or white asbestos—may, in its pure form, have no health effects more significant than from other mineral dusts that are benign when pure, such as limestone and talc. The key word here, it turns out, is *pure*; Zimbabwe's chrysotile is exceptionally so.

There are five dangerous asbestos types, all falling into a mineral category called amphiboles. Two of these used to be mined commercially in South Africa, crocidolite (blue asbestos) and amosite (brown asbestos). Of the other three—actinolite, anthophyllite and tremolite—it is necessary to take note only of the last.

Until about 1930, chrysotile asbestos was the only sort in general use in the United States. The records of the relevant worker's union (The International Association of Heat and Frost Insulators and Asbestos Workers, whose members are known as 'Insulators') show that the correlation with mortality due to asbestos diseases only developed after the introduction of the amphibole, amosite (brown asbestos), in their products.[129] This occurred from about 1930 through to 1934, but the slow onset of the diseases involved meant that it was not until well after the Second World War that warning bells began to sound.

All poor together

Following the alarm signals from South Africa, an investigation of the UK records showed a dramatic rise in the incidence of mesothelioma occurring for workers who came into contact with 'asbestos' after 1940. One example gained some notoriety from a 'Panorama' television exposé called 'A Day in the Death of Alice'. Alice was one of a number of British women who contracted mesothelioma and died after packing gas masks with crocidolite pads during the Second World War; by 1978 fifty had died or were facing death in this form.[130] Amongst their families there had been a number of mesothelioma fatalities as well, thought to be caused by the dust being brought home from the workplace in clothing. It was a disease that seemed to be independent of smoking, but related to the amount inhaled from the time of first exposure to the fibres.

There was an anomaly here, which the 'Panorama' programme did not mention. This was that another group of women, 757 of them, who were packing gas masks with chrysotile pads, showed no ill effects.

Nonetheless, chrysotile was still implicated: the percentage of miners and textile operatives working with this type who were dying of mesothelioma was found to range up to 2%.

By the 1970s the issue had moved beyond the scientific sphere. The outcry over the discovery of relationships between 'asbestos' and diseases brought the question into the political arena, and from about 1964 onwards regulatory decisions were largely based on a need for administrations to be seen to be doing something about the hazard.

The emotional pressures were immense, and the responses were in proportion. From 1940 to 1973 schools across the United States were *required* to have asbestos insulation as a fire safety measure. Then, in an abrupt reversal, the Environmental Protection Agency banned the installation of asbestos in schools in 1973. The Asbestos Hazard Emergency Response Act of 1986 required all private and public schools to inspect for asbestos, develop asbestos management plans, and implement appropriate actions. By 1990 it was estimated that the cost of such abatement work was over $6 billion.

Something in the air

In 1970 the U.S. Congress enacted the Clean Air Act. A provision of the act allowed the Environmental Protection Agency to designate certain substances as 'hazardous air pollutants'. Asbestos was one of the three such substances on the first list of 31 March 1981; no attempt was made to distinguish between the various types.

The rise of public opinion against the use of asbestos had disastrous effects on the producers. Between 1973 and 1982 sales of asbestos in the U.S. and Europe declined by 700,000 tonnes, or about 30% of exported production. This incrimination of asbestos as a health hazard and carcinogenic agent led to the withdrawal of the world's largest producer and manufacturer, the Johns-Manville Corporation, from the industry. At the time it faced 16,500 claims for asbestos related illness, and the company was advised that there could be as many as 52,000 cases, with the risk of damages exceeding $2 billion. Manville had purchased significant amounts of insurance cover, but the insurers refused to pay claims, saying that Manville failed to inform them about the dangers of asbestos in the 1930s, thus making the policies void. In desperation, the company filed for bankruptcy under Chapter 11 of the Bankruptcy Act, saying that the company had been 'overwhelmed' by legal actions. The Chapter 11 filing automatically stopped all lawsuits.

The process so started finally culminated in Manville's sale of its Jeffrey mine in Canada—the biggest asbestos operation in the world—to JM Asbestos Inc., a privately owned company formed specifically for this purpose.

By 1994 the onslaught of asbestos claims in the United States had caused gridlock in the courts. More than 100,000 cases were pending with many states still dealing with claims filed in 1980. At that stage the U.S. asbestos industry had paid out in excess of $10 billion in compensation and very few companies continued to operate. A study was published by the Yale School of Organization and Management in 1992 in which the writer, Paul MacAvoy, predicted that there would be 200,000 asbestos-related deaths in the United States over the next 25 years at a cost to asbestos manufacturers and their insurers of $50 billion.

Indeed, although Britain's death rate per head of population

All poor together

from asbestos is about four times higher than that of the United States (because of the wider use in Britain of amphiboles), American asbestos victims are five times more likely to sue for damages because of the more active role played by doctors, lawyers and trade unions there in advising sufferers of their rights.

Attempts have been made to find alternative products to asbestos in virtually all its areas of application (of which there are about 3,000). These efforts have been reasonably successful, as judged by the progressive decline in the consumption of asbestos in the U.S. from 672,000 tonnes in 1977 to only 83,000 tonnes in 1988. But, given the time it takes for cancerous tumours to develop in the lungs, we could be in for another shock in twenty or so years time, when it turns out that one or more of these substitutes are also carcinogenic.

In Zimbabwe, the output from the mines also fell as the markets dwindled, from over 270,000 tonnes in 1976 to 165,000 tonnes in 1996. Employment dropped in tandem; 6,000 workers were laid off in the same period. In 1994 the owners at that time, the United Kingdom's T & N plc (previously Turner and Newall Ltd), were forced to make a £140-million ($224-million) provision for asbestos-related disease claims. This severely affected profitability; in 1996 T & N set aside £515 million for future asbestos-related disease claims, contributing to a pre-tax loss of £388 million for the year. In 1997, with some relief, they managed to dispose of their Zimbabwe asbestos mines and factories to a locally owned group.*

Which brings us back to the health anomaly of these mines.

---

* The hazards of corporate hubris were noted by C. Northcote Parkinson, he of Parkinson's Law fame ('work expands to fill the time available for its completion'). He observed that organizations that have reached the stage of building themselves a grandiose, albeit carefully designed, head office invariably start to deteriorate from that point ('a perfection of planned layout is achieved only by institutions on the point of collapse'). Similarly, perhaps, with the publication of a book surveying the triumphal growth of an institution. Turner and Newall published such a volume just as the deluge was about to break. It was called *Turner and Newall Limited: the First Fifty Years 1920–1970*, and in the foreword the then chairman wrote—'The history which you are about to read represents not merely the end of a story but the beginning of another, more exciting and more challenging still.'

The problem is not that Zimbabwe chrysotile appears to be safe; it is that the health statistics from the Canadian chrysotile mines are not as clear cut as those from Zimbabwe. Could it be that there is something wrong with Zimbabwe's data, or its interpretation? It could not be ruled out, as a well-respected local doctor noted in 1983—

> Either the Zimbabwe asbestos industry is relatively free from risk to health, as is suggested by the foregoing *prima facie* evidence, or the hazards are greater than we believe, hidden because we have not looked hard enough.[131]

Certainly, in the Canadian chrysotile mining areas the incidence of mesotheliomas (but not of lung cancer) is above normal. Amongst nearly 10,000 Quebec miners, 31 cases had been found up to 1992.[132] The cause of this phenomenon seemed clear. A 1996 paper[133] analyzed the same Canadian data to assess the apparent potency of chrysotile in the causation of mesothelioma. It concluded with the broad and unequivocal accusation that:

> Chrysotile asbestos is by far the main contributor to pleural mesothelioma causation in the U.S. and other countries in which it has been the predominant fibre type. Crocidolite may be 2–4 times more potent, but there is no valid evidence that amosite is more potent than chrysotile.

There was more. A 1980 review of information on the different pathogenic effects of the different forms of asbestos on laboratory animals[134] stated firmly that:

> Almost all the (in vivo) injection studies showed that chrysotile was at least as carcinogenic as the amphibole dusts and usually more so ... All inhalation studies where tumours developed have found chrysotile more effective in producing bronchial carcinomas than amphibole dusts, and chrysotile also produced at least as many mesotheliomas ... Without exception a series of studies examining the haemolytic effects of asbestos have reported that while chrysotile is highly haemolytic, the amphiboles show very little effect.

But if chrysotile was really the villain in Canadian mines, where were the asbestosis, lung cancer and mesothelioma cases in Zimbabwe's mines?

In 1997 came an explanation, in the form of two papers. One of these[135] examined the employment of 11,000 men born between 1891 and 1920 (in statistical sampling terms, a 'cohort') who had worked on the Quebec chrysotile mines. Of the 8,009 who died between 1936 and 1993, 38 probably died from mesothelioma. When the records of these workers were examined it was found that there were geographical differences in the scatter of infections, with those employed in the five mines in the central area of the Quebec asbestos belt (the Thetford mines) being 2.5 times as likely to be affected as those in the outer area (the mines around the town of Asbestos).

In statistical terms this difference is highly significant. As the authors said in the paper:

> When this cohort was set up (in 1965) in response to international requests for study of the effects on human health of exposure to pure chrysotile and other asbestos fibre types, no serious thought was given to the possibility that the Quebec ore body might contain mineral fibres other than chrysotile ... Soon, however, evidence that the situation was not so simple began to unfold.

In 1979 a study of the radiological evidence showed lung shadows to be much more common in men who had worked on the central group of mines. To quote again—

> A few years later, studies in lung tissue at autopsy from miners and millers in the Thetford (i.e., central) area revealed the unexpected finding that, despite overwhelming exposure to chrysotile, amphibole fibres of the tremolite series were present in similar or greater concentration.

Tremolite, it will be remembered, is one of the rarer amphiboles, and one that has never been mined intentionally. A mineralogical investigation showed that the source of this mineral was dykes (underground volcanic intrusions) that had pushed into or near the chrysotile ore body. In one area these dykes were numerous and of irregular shape, meaning that more waste material from them containing tremolite would be liable to be mined along with the ore than in the outer area of the belt. On investigation it was found that tremolite contamination of chrysotile was four times the level of the outer mines. So here was a vital piece of evidence that exonerated pure chrysotile fibre from any especially malignant property.

The second paper, published at the same time, went further.[136] Tremolite had already been shown to be responsible for a very high incidence of the full range of fibre-related diseases (mesothelioma, lung cancer and pulmonary fibrosis) amongst American vermiculite miners.[137,138] Now, using the information on lung cancer amongst the same cohort of 11,000 men, a correlation with lung cancer and the incidence of tremolite was also discovered for Canadian chrysotile mines.

These papers were not the first indications that minor amphibole minerals were causing the trouble. In 1987 a chrysotile mine in Cyprus was thought to be the cause of a local rash of mesotheliomas, until a post mortem showed tremolite in the lungs, whereupon this mineral was found to be widespread in the mine.[139] In 1990 it was concluded that a rare variety, called balangeroite after the chrysotile mine in Italy where it was found, could be responsible for the three mesothelioma cases found amongst miners there.[140] In 1999 a death from mesothelioma was recorded from a worker at the Cullinan Diamond Mine in South Africa; his lungs contained tremolite.[141]

With the tremolite discovery much else fell into place: the troubling anomalies of mesothelioma cases in importing countries where Canadian chrysotile had been used; the discrepancies in animal studies, which are now thought to have employed similarly impure chrysotile;[142] the absence of mesotheliomas amongst the Zimbabwe chrysotile miners.

The discovery has probably come too late for the 6,000 workers remaining on the Zimbabwe mines and the 5,000 others who work in the downstream factories there that make fibre cement products. The impetus given by the horrors arising from the amphiboles has led to a world-wide wave of indiscriminate banning. The Zimbabwe government, faced with another blow to an economy that has a 50% or so unemployment rate, has appealed to the European Union to fund an independent survey of the medical evidence on Zimbabwe's chrysotile, but the EU's Commissioner for Development at the time, Professor Pinheiro, brushed this aside—

> I am not aware of any counter evidence to dispute the fact that all types of asbestos may cause lung cancer and mesothelioma, a form of cancer of the lining of the lungs. We have taken due regard of the views of the independent Scientific Committee on Toxicity, Ecotoxicity and the Environment who have stated that no safe threshold for chrysotile asbestos can be identified.[143]

Not all of Africa's woes are brought on by herself.*

---

* The EU has, however, indicated that it might assist with the 'restructuring' of Zimbabwe's chrysotile industry. At the present time about U.S.$50,000 of investment is needed for each new industrial job in Zimbabwe, so 11,000 of them could cost the EU taxpayers $550 million. An independent medical survey would be a fraction of this, but the EU has evidently decided to substitute money for political courage.

Chapter Nineteen

NEVER ENOUGH

From the red gold keep thy finger;
Vacant heart and hand, and eye,
Easy live and quiet die.
—Sir Walter Scott, *The Bride of Lammermoor.*

That day the fallout from a distant harmattan descended upon the camp like a hot grey mist. It was December in the dry, sparse bush of Burkina Faso, and the little gold plant I had been commissioning was halted because of a broken pump. Towards evening a breeze cleared the atmosphere, but the sky remained opaque and sullen and all distances were shrouded in haze.

Abdul Karim and I had taken advantage of the shutdown to investigate a hill about four kilometres from the camp. From the scars we could see near the summit it appeared that the *orpailleurs*—the ubiquitous local gold miners—had been digging away at a discovery up there.

But when we pulled ourselves gasping to the top we found that the marks had been caused by the erosion of the dark laterite that caps all soils out here. There was bauxite and kaolinized schist present, as well as the odd piece of quartz, but no reefs. But at least for a time I had escaped from the vast slum of the *orpaillage*. Far below us the sprawl of flimsy huts that composed the camp was almost hidden by the haze and the gathering dusk.

Abdul started to talk to me about money, in the Frenglish lingua franca we used. He muttered steadily away, showing no emotion; he was not expecting me to help, he just wanted me to know that he had his problems too. By Burkina Faso standards he was well off; his earnings were about 185,000 CFA francs a month, equivalent at that time to about U.S.$370.

But it was not enough, he complained to me. It never is and never will be, I assured him. I had no say in what he earned, so I could afford to be philosophical. Abdul replied by listing his obligations. He was supporting two brothers through a local high

All poor together

school, making a contribution to his father every month, helping his mother out (his father had two wives and nine children; Abdul's mother was the junior wife) and at the same time saving to buy twenty goats in order to purchase his own wife from an avaricious father. The only asset he didn't have to save for was access to land; Abdul's father had got his local chief to allocate him a couple of hectares.

All told, Abdul's income shrank to a net of about 30,000 CFA francs after all his obligations had been met. He was almost as poor as the ragged workers digging away at the mine.

Three months before I had been 5,000 kilometres away, in southern Zimbabwe. With a local geologist I was seeking a long-abandoned gold mine lost somewhere in the endless mopani-tree bush of southern Matabeleland. My colleague, Gilbert Mutasa, was also earning a fair salary by Zimbabwe standards, with a take-home salary of about 10,000 Zimbabwe dollars a month (about U.S.$600 then).

Yet he too was struggling to survive on a fraction of that. He too was paying for the education of a couple of siblings. He too was accumulating livestock—cattle in his case—for a bride price. He too was a child of a polygamous marriage.

So it goes all over Africa. Culturally you might expect Abdul and Gilbert to have nothing in common. Abdul's native tongue was Mossi, a negroid language which has almost no linguistic connection with Gilbert's Chishona, a bantu language. Abdul is a Muslim, Gilbert a (nominal) Christian. Abdul's European language was French, Gilbert's English. Yet their cultural basics are identical—communal land ownership, polygamy, a bride price denominated in livestock and responsibility for the extended family that arises from this environment.

It came to me, sitting up there, that the homogeneity of Africa's culture is amazing. In the bush of all the eighteen countries I have worked in, the people struggle in a thick net of common obligations. Whether it is in the dry scrub of the Sahel, in the forests of Ghana, in the cassava fields around Lake Victoria or among the granite whalebacks of Zimbabwe, the inhabitants—the men—all have the same ambitions and obligations: to acquire livestock, to buy a wife, to educate a cousin, to support a parent, to have more children. It all needed much more money than a man could ever make. Abdul stared out over the barren land, shaking his head.

The dusk thickened and we scrambled back down over the decomposing slopes and their hindering thorn trees. I was living in that *orpaillage* below us, an artisanal miners' camp called Dama. (*Orpaillage* means a working place for gold washers; I never heard the locals use the word, they simply called it *le site*.) Here I was starting up a tiny retreatment operation for Bernard, my client, a plant that used cyanide to extract the gold from the *rejets* there. These were the residues left after a long succession of hopefuls, men at first, then women, had panned and repanned the ore that the miners had brought to the surface. Bernard, brave and optimistic, like all investors in gold, said that there were 6,000 tonnes of this material. I thought more like 4,000, but it was running at 15 grams of gold—half an ounce—to every tonne, ten times the amount that a similar plant might treat profitably in Zimbabwe.

As we descended I reflected that, culture aside, there was another striking similarity I had found across Africa. The geology here at Dama was identical to that in Tanzania around the southern part of Lake Victoria, nearly 4,000 kilometres away. It was hydrothermal, archaean, with the gold in quartz reefs and silicified shear zones, capped with laterite. In both countries, therefore, the reefs had been largely hidden, were discovered by chance and their exploitation had been by an unplanned rush of incomers, many with some experience in their trade. In both countries the phenomenon of artisanal gold mining had only got under way after the gold price went up in 1971.

These reefs and shears were, on the whole, too wide to be mined by rows of narrow circular one-man pits, so a succession of wider, square shafts had been dug by groups along the strike, anything from ten to a hundred of them, with centres between three metres and five metres apart.

The geology, then, had enforced the same mining method—pitting—in both Tanzania and Burkina Faso. Tanzania, where I first saw it, was rather more technically advanced, with a widespread use of pumps, crude windlasses, ventilation fans (made largely out of bicycle parts and irrigation tubing) and with mercury almost invariably used to purify the panned concentrates. But now, five years after I had first seen Burkinabé artisanal miners in action, windlasses and mercury were an accepted part of the technology used (no ventilation as yet) and there were many more pumps. There was also—something I had not seen in Tanzania yet—grain mills being used to regrind the hand-crushed quartz.

Further, not only was the mining method similar but the nature of the ownership and even the organization were the same, down to the arrangements for the division of the spoils.

In both countries the ownership of the find is not 'finder's keepers' as it should be. Applications have to be made in person at a remote ministry (postal applications are too easy to ignore) and public information on the law is limited, with no readily available translation in local languages (this is more important in Swahili-speaking Tanzania than for Francophone West Africa, where the proliferation of languages has meant that French is almost universal). As a result, for both countries the acquisition of mineral title requires resources of literacy, time and money that the average discoverer does not possess. Hence the discovery, if it is any good, is usually usurped by a 'big man' in politics or business.

Again, in both countries the individual pits were excavated by groups of between four and fifteen miners. In both they dug down to the water table, where there was a certain amount of widening out to search for supergene enrichment. Sub-contractors (*chefs des puits* in Burkina, 'pitholders' in Tanzania) assembled these groups, which were often bound by ethnic ties, but in many cases from the comradeship generated by working on previous diggings together. In Tanzania the claim-holder, in Burkina Faso the *chef de orpaillage*, takes half of the value of the gold produced (in both the gold buyers, as well as the pits themselves, are carefully watched by agents of the owner). In turn the *chef de puit*/pitholder takes half of what is left for himself. Of the 25% remaining, some has to be given to the women crushing the rock that is sent to the surface. As the ladies with their *mortiers* work on a fixed price (about $5 per sack of ore usually), little is left for the workers unless a high-grade area is found. They got food, certainly, and as we have heard they steal a fair amount through the *prizi* system. However, their average income in both countries is probably only between $10 and $40 a month.

I was reflecting on this illustration of how a common geological environment has led to the evolution of a common social structure when we arrived back at the refuse-ridden outskirts of the Dama *orpaillage*. In the dusk, fires glimmered outside the makeshift huts, and the shadowy inhabitants shuffled about us, muttering *'b'soir, b'soir'* as we passed. I thanked Abdul and made my way past the thronged, murmuring shacks towards the steady gleam of the lamp at my own quarters.

'They live in heaps', T. E. Lawrence had said of the Bedouins' lack of private lives; here at Dama it was the same. Goats, donkeys, sheep, chickens and flocks of tame white guinea fowl ranged everywhere. Ragged children of all ages, worn-looking women, young men, old men, ancient men, drifting steadily about in ones and twos, washing, spitting, praying, faces turning to watch as I drink tea or eat supper on my 'verandah', a patch of corrugated iron shade. A little bit of begging from some small children, but nothing pushy, just trying it on.

At the plant the afternoon shift was coming off. These workers, the young men who are doing the shovelling and wheelbarrowing, were wilting a little by this stage. In the coal mine in Zimbabwe where I once worked, new labourers took about two months to attain the performance of seasoned miners; it took that long to build their muscles up to the strength needed. Men who live in the bush usually do not eat well enough for protracted heavy manual labour in scorching heat and it is the women who do most of that anyway.

These Burkinabé had names like Tapsoba, Mamadou and Omadou, and wore the cast-off clothing of America, sent out in bales and sold cheaply in every souk and marketplace on the continent, crowding out the sales of the frail indigenous textile enterprises. They had checked and striped trousers (no shorts) and jeans whose bell-bottoms dated them to the Beatles era, and

All poor together

grubby T-shirts offering a wildly irrelevant range of slogans—'Somebody in PANAMA loves me', 'D.A.R.E. to keep kids off drugs','OAKLAND Athletics 1972', 'BUGLE BOY'. I was worried for their safety after I left, when they would get too casual about handling cyanide. I adjured Abdul to keep reminding them of its dangers and how to use the antidote.

It had taken three days to get here from Harare. A flight down to Johannesburg, one up to Abidjan, one onward to Ouagadougou and then five hours in a Land Cruiser to the *site*. I had with me three cast-iron slag moulds, cone-shaped, for the gold smelt, and expected to be challenged at every X-ray (I was to go through four, including two at Johannesburg), but, oddly, no query was made.

Air Afrique is flexible with its time-tables; it has to be because of the unpredictable character of its destinations. At the time, trouble in both the Congos meant that the aircraft had arrived at, and left, Johannesburg an hour early. A new stop for me would have been Point Noire in the Congo (Brazzaville) but this was now held by one of the factions competing for the government (as it happened the democratically elected one) so we did not stop at all on the way to Abidjan, overflying Kinshasa, Brazzaville and Pointe Noire.

In retrospect it is easy to forget the nagging uncertainties that surround you all the time during this sort of travel: the impossibility of getting confirmed seats, making each transfer a fraught step into the unknown; the forgetting of some important piece of paper (in this case my yellow fever certificate); the need to comprehend roaring, incoherent flight announcements that leave everybody in the departure lounge (lounge!), the French-speakers as much as the rest, looking around blankly at each other; even something as mundane as trying to ensure that you get on early enough so that you can get your hand luggage 'stowed in the lockers above your head' instead of 'under the seat in front of you'(which can mean hours of cramped knees) looms hugely over your peace of mind.

Luckily Bernard had arranged a *facilitateur* to help me through the obstacle course of arrivals at Ouagadougou. That was nice, but not necessary; I had done it single-handed before. Still, this time I was grateful; with this fellow in attendance the man in the *santé* cubicle just waved me through. Even the customs scrabble-and-peer was perfunctory.

At the exit I was reminded that Burkina Faso is one of the countries scrabbling for the bottom position in the poverty ladder. Here no less than five members of the crowd made a plunge for my *valise*. Not to steal it, but to carry it for, they hoped, a tip. My *facilitateur* selected one, and we all walked through the warm languid dusk to Bernard's car.

Airport terminals, international-class hotels, night clubs, up-market shops, golf clubs—anywhere that the people with real money in their pockets go—have become areas to be discreetly patrolled by these young (and not so young) men. Should an opportunity for any service, however trivial, appear, then they are suddenly there, underfoot. It was one of the best things about the *orpaillage*; I might be under constant surveillance, but there was no importuning.

This was a dying camp. Many of the huts were empty and most of the miners who remained were walking daily to a new discovery about four kilometres away. The main centre of earlier activity, the courtyard or plaza or piazza or town square—an open space, anyway, bordered with thatched lean-to's where the crushing and panning had gone on—was inhabited only by women and children now, repanning for the sixth or seventh time the piled *rejets* that occupied the centre of the space, in the hope of finding enough left-over grains of gold to live on. A couple of pits were still being worked, and the rythmic thump of pestle and mortar (a wooden mortar, a lorry half-shaft for a pestle) could be heard intermittently. At a new discovery there would be a continuous rumbling barrage of noise from scores of these crude crushing devices.

It was the middle of the dry season, but the water supply that Bernard had found (at the third attempt, a sluggish, brackish borehole two kilometres away) was a huge boon. In this desperate land, when the rains come each maize plant is grown in an individual hole pecked through the surface laterite. Now, with water, the maize has already been planted and women and children were using the big plastic tea-pots that are sold everywhere in West Africa as watering cans. The contrast with the endless, dense stands of maize on Zimbabwe's commercial farms could not be greater.

My days were busy with the little triumphs and disappointments of starting up a plant novel to the country—to the region, even—which had been built under my guidance at very long range. The system was the same rudimentary one used, years ago now, at Buhemba; a series of shallow tanks, each with the floor covered by a layer of stones, then sand and finally sacking. This allowed the gold-bearing cyanide solution to drain through the *rejets* that had been shovelled in to an outlet in the base, while the *rejets* themselves were retained. A long rectangular tank, divided into compartments that had been filled with zinc shavings, was to precipitate the gold from the solution. No electricity, just wheelbarrows and shovels and a little petrol pump to pump the solution around. Control, such as it was, used an ancient method known as the Purple of Cassius test to detect if the solutions had gold in them. As might be guessed, this produces a strongly coloured complex (of gold and tin) to indicate the least trace of gold present; it is said to resemble the imperial purple of ancient Rome.

So the plant could hardly be simpler. Nonetheless there were crises: the wrong pump, dirty sand, missing chemicals, choked valves (*robinets*, what a pleasant name) and some brow-furrowing moments over the contrary results from the purple of Cassius tests—and my diary is largely an account of these. But there were lighter moments, such as the business of the face masks. The cyanide had arrived as a powder which was liable to fly around dangerously (I had ordered pellets). Bernard's drivers were wearing simple face masks made of sky-blue filter cloth, the colour of the company logo and overalls; a nice touch. So I started a wild goose chase to find more of these for the plant workers. After a lot of confusion (my execrable French did not help), I discovered that these were not the real thing at all; the drivers had found some old Air France eye shades, coincidentally the same colour as that adopted by the *Societé des Mines de Burkina* and were using them as (presumably rather impervious) dust masks.

The real thing, when located, were disposables made out of

tough paper. Abdul issued them to the workers, some of whom took to wearing them on their foreheads. Those that did wear them properly appeared, at a distance, all to have the palaeanthropic mouth of Bart Simpson's father.

I was referred to, and addressed as, *Le Patron*, which was nice. I had a bed with a mosquito net, a primitive bathroom, and a cook/valet who was able to make the tiny tough local chickens almost edible. I had beer; the stuff from the *Brasseries de Burkina* was quite drinkable in the heat. I was very busy and there were no distractions: the nearest telephone was at a *gendarmerie* forty kilometres and one hour's drive away. At some stage during those weeks I realized that I was very happy as well.

The main worry was the cyanide. The day after we started up, Abdul found a dead sparrow at one of the tanks that had died from cyanide poisoning. He took it away to show the local *proprietaire des moutons* what cyanide would do to his goats if he let them wander in. That day the box containing the cyanide antidote was put in and warning notices erected. An old man was deputed to be gate guard, and he quickly acquired the status of a chair to sit on. The morning *equipe* were all sent off to cut thorn-tree branches to reinforce the fence with a *scherm* (French speakers should not turn to their dictionaries; for lack of anything better I used this Afrikaans word to describe the palisade of thorns we erected, and it was rapidly adopted there).

A flock of laughing doves descended on the tanks at midday while the *equipe* was lunching; I drove them off, but one pair dipped their beaks in the solution on top of one of the tanks. They remained sitting on the side as I approached, then the female flew awkwardly away but I picked up the male, who scrambled out of my hands only to flutter to the ground. I launched him back into the air towards the fence, but he hit a pole and crashed down. I tried again but he nose-dived into the thorns on the other side. There was nothing to be done.

From then on, every day saw a gathering-in of dead birds, mainly what bird-watchers call LBJs (little brown jobs). They were carried off triumphantly by the workers for food. The tiny amounts of cyanide needed to kill these birds would have no effect on their consumers, particulary after they had been cooked. One evening the old man at the gate showed me his haul, about 30, half-and-half pigeons and LBJs. Whether out of sympathy or irritation with having to gather them up, he has been running water into one of the drainage trenches for them to use.

This was the most distressing part of the job, but there were more potent worries. The tanks were draining far too slowly; whatever the panners had done to the *rejets*, they had given them a glutinous consistency when wet that defied quick percolation. The lime supplied turned out to be unslaked, requiring careful handling if nobody was to get burnt until we had slaked the whole lot. We had tremendous problems getting the clear solutions necessary (remember the 'gin clear' of Buckreef?). Eventually we stopped everything and rebuilt the sand filter beds in the clarifiers.

So far I had not taken any photographs, out of a superstitious fear that they would mock me if the whole enterprise collapsed. But, after ten days of disruptions and alarms, the solutions had become clear, the zincs were giving off little bubbles of hydrogen as the gold precipitated on them and the cyanide solution was percolating through the *rejets*, albeit at a leisurely rate. The colour test was still giving odd results, but I decided to chance a little hubris, and started snapping. Most of the pictures had Abdul in them. He was given to striking stern poses, particulary in front of the sweating labourers; they were to impress his 20-goat intended. As for me, well, I may indeed have been rather premature.

Abdul had gained what English he had while working for a Canadian company with a concession here. Like many such, it had seen its shares collapse in the aftermath of the Bre-X scandal, and the directors had abandoned their exploration in Burkina Faso. They were disappointed with Dama anyway, it seemed.

Why was explained one evening when Abdul and I went down some of the pits, so that I could examine the reefs *in situ*. The first pit was, he said, 32 metres deep. He had a rope, but it was not long enough, so we went down some shallower ones, about seven metres deep, torches stuffed into our belts. In the dank-smelling bottoms of these we cast about for ore, crawling along drives and chipping at the walls. We found reefs with good-looking dark and mottled quartz, with much promising intergrowth between the quartz and the sheared andesitic country rock. This accounted for the Canadian interest; they were looking for big, shallow well-weathered ore-bodies that could be open-casted and heap-leached. Here they had initially found something rather like that, because near the surface the gold-bearing andesite had weathered into clay and the values had been washed out all around the reef. But not enough of it; the weathering was only about 30 metres and the gold was not dispersed far enough from the quartz to make a decent reserve. All that was left of their efforts was a long trench that, mysteriously, ran away from the main strike of the *orpaillage*.

Bernard had left me a copy of their report. It didn't take matters much further forward except to recommend further trenching, and to acknowledge that the gold was in the reefs and effectively nowhere else. The obvious thing now was to drill the main strike of the *orpaillage* at depth, but nowhere did it recommend this. (Readers will know why; there's many a good prospect ruined by drilling.)

Two weeks after we had got the plant going it was clean-up day. We stopped the flow of solution, took off the locked covers of the zinc boxes and started draining the compartments. Bernard arrived from Ouagadougou at that point, and I pointed out to him the curious yellow colour on the zinc shavings. Could it be gold? I told him that finely divided gold is black and that is the appearance they should have had. But then again I had never dealt with such a high-grade material. So, perhaps it could be.

We carefully scrubbed the hanks of shavings under running water so that the gold slime washed off into buckets, and then replaced them in the compartments. It was a slow business, and it was not until seven in the evening that we set off back to Ouagadougou, the slime in two buckets held by me on the back seat of the Land Cruiser, where I tried to prevent them from spilling. We drove slowly through the darkness over the laterite tracks, and it was one in the morning before I got to sleep; Ouaga seemed abrasively noisy after the murmurous peace of Dama.

The next day — a Sunday — we drove with the precious buckets across the sprawling city to *le jardin de M. Do*. Konaté

Do's *jardin* was actually a fly-ridden earth yard, with not a piece of greenery in it and largely occupied by broken-down cars, a working lorry and some sheep (which were chased out for the occasion). Here I was to dissolve the non-gold components of the black slop in the buckets with sulphuric acid, wash the residue with hot water and then roast it to a calcine, oxidizing any base metal that was left.

I realized when I looked in the buckets in the bright daylight that there were very few small pieces of zinc present. There should have been a lot of these, as the zinc goes into solution when the gold (and other metals) precipitates, so the shavings tend to break up as a result. It turned out that in the darkness these 'shorts' had been passed from hand to hand back into the zinc boxes, not the buckets, which meant that much of the gold might still be at Dama.

But there was nothing for it but to press on. At least with no 'shorts' there would be a very low acid consumption. As at Dama, improvisation was rife. I needed some calico as a filter cloth; there was none and so M. Do's *mouchoir* had to serve. The lime to coat the surface of the calcining tray had been left behind so we used wood ash from the fires heating the wash water.

Alarmingly there did not seem to be much calcine left after roasting. Konaté Do said maybe a hundred grams, I said heavens, no, more like two hundred. But only two hundred! Where had all the gold gone? That yellow stuff I was so excited about must have been iron, or kaolin stained with iron. Bernard and I returned to his house and spent a gloomy evening together, although it got better after I asked Bernard to open a bottle of wine.

But I spent much of the night in thought. We were due to smelt the next evening at five. The colour tests that had given us so much trouble may have been wrong after all. Perhaps, horrors, we had not been leaching gold at all. There was a lab in Ouaga that could do rapid solution analyses; should I go back to Dama to get a range of samples from the various stages of the leach? There was a very good reason why I should not: my air ticket had me flying the day after tomorrow. If I missed that flight I would either have to hang around in Ouaga for four days until the next or launch myself into the blue on the first flight to Abidjan, in the hope that I could get my ticket endorsed there to a carrier other than Air Afrique.

I awoke with my mind made up. I owed it to Bernard to see this through. Early in the morning I set off with Bernard's driver to drive through the grey, featureless bush all the way back to Dama. We had hoped to return the same evening but in the event Abdul's arrangements had meant that some of the solutions needed would only be available next morning, so we spent the night there. Luckily it was hot; there was nothing but a bare mattress to sleep on. I had not even brought a toothbrush. There was no chance now of flying out from Ouaga tomorrow.

I got up at five in the morning and wandered about the darkened plant, taking samples. We set off back to Ouaga an hour later and by midday we had the results. These were very strange, absolutely the reverse of the normal order of events, with the gold contents increasing day by day, not going down. The grades, though, were very, very high; even the barren solution after the zincs was three grams a tonne, way above most pregnant solutions in Zimbabwe mines. Still, gold was coming into solution and being precipitated, and when Abdul telephoned from the *gendarmerie* that afternoon I gave him a list of changes to make to the process. Or rather Bernard did; it was too important a matter to allow my French to interfere with. As they talked I heard my missed flight go roaring over, on its way to Abidjan.

Now for the smelt. We drove with our modest packet of grey calcine across to a Monsieur Charles, who was a gold dealer and had the requisite crucible furnace. M. Charles, his professional pride aroused, had a polite but intense argument with me about the fluxes needed, but in the end he agreed to use my recipe. (And another mystery solved. One of the fluxes needed was silica, which Bernard had complained was incredibly expensive. I had the uneasy feeling that whatever it was that was so pricey, it was not what was wanted, and had brought some crushed quartz from Dama for the purpose. It turned out that Bernard had acquired nearly a thousand dollar's worth of beads of silica gel, the stuff that is found in tiny packets to remove moisture from stored optical and electronic equipment.)

In an atmosphere of strained competition I weighed out the

All poor together

borax and silica, mixed them thoroughly with the calcine and divided the mixture amongst three small plumbago crucibles. (Plumbago in this context is not a flower: it is a mixture of graphite and clay.) M. Charles then started the air blower and while it roared away went through an exciting business with matches, propane gas and diesel oil to get the furnace lit. With our eyebrows only slightly singed, we lowered a crucible into the fierce flames and waited. The moulds that had featured on so many airport security screens were oiled and ready on the floor next to the furnace. It was immensely hot.

Speech was impossible, but I was determined to do it all myself from now on. I swathed myself in leather gauntlets and gaiters, a leather apron and a face mask of clear heat-resistant plastic. Then with long tongs I removed the cover from the furnace. In the incandescent heart of the crucible a molten surface flickered and swayed. It was done.

I grasped the crucible fiercely in the tongs, pulled it carefully out of the furnace, and tipped it slowly over the mould. The glowing contents ran out into it with a poppling noise as the oil flashed off, but lumpily, sluggishly. Damn! The flux was wrong. But there was nothing for it now but to wait and see what was in the mould after it had cooled. There was a sudden silence when M. Charles switched off the blower; his face impassive with suppressed justification. I realized that I was drenched in sweat, and pulled off the gloves and mask.

The silence held as we watched the contents of the mould darken. Small cracking noises came from it as the slag contracted. After five minutes I could wait no longer, and with the tongs up-ended the mould. The black cone inside fell cleanly out and lay crackling on the concrete. There should be a button of gold at its tip. I could see none. Bernard's lips pursed.

M. Charles went to a sink next door and filled it with water. The hot thing on the floor was now noisily breaking into pieces as it cooled, and I swept it all on to a metal plate with a brush whose hairs singed as I did so. M. Charles tipped the contents into the water, amidst a great hissing, and then bravely felt around in the sink. He came up with a solid, steaming, blackened lump, and, bouncing it in his hand to avoid being burnt, put it back on the plate. It seemed to be only slag. Memories of a white-robed Arab looking distastefully at a similar lump in far-away Buhemba rushed back. 'It looks like copper', he had said.

I gazed more closely at the hot thing and pushed it about cautiously with a finger nail. For slag it seemed to be very heavy, and it was not cracking. I took a hammer and tapped it. The coating of slag fell away to reveal a rich yellow bullion. 'L'or!' Bernard exclaimed with huge relief.

In the end the total weight of bullion was about 300 grams. Using water to estimate the density of the buttons (now combined into a little yellow cone of satisfying heft) we decided we had recovered about eight ounces of pure gold and there would be more, perhaps an ounce, to get out of that first sticky slag. Not marvellous (we had hoped for fifteen ounces or so), but a whole lot better than my first fears. As for the rest of the gold, the samples I had taken in the early-morning darkness of that day had already told us that most of it was still in solution. Now Bernard was beaming and I was suddenly very tired.

· · · ·

Of course, the Dama story does not end there, it just gets less interesting. It is enough to say that in the end I got back home only three days late and only a few hundred dollars poorer (after I received my ticket refund from Air Afrique six months later).

In the meantime I was given ample time to meditate on the revelation that had come at the top of the hill near Dama: that the far-flung inhabitants of sub-Saharan Africa all live in the same culture. Indeed it may be said that my Damascene conversion occurred while a long-time resident of the noisy, confused departure lounge of the *Aeroport Houphet Boigny* in Abidjan. But to tell the whole story of this awakening it is necessary to go to its beginning, which had occurred a few years before, in the foothills of the Argentinean Andes.

Chapter Twenty

IDA WHO?

> 'The observable behaviour of post-independence Africans could not, therefore, be seen to be modelled after that of a white man's role. We had to be seen to be jettisoning everything that our ostensible antagonists stood for. But—big problem, what to replace the role model with?'
> —Dr Nkosana Moyo, Managing Director of Batanai Capital Finance, Zimbabwe 1999.

The suburbs of Buenos Aires straggle endlessly westward. On the outskirts there were sprawling slums of flimsy wooden shacks, with a tangle of electricity lines drooping down from illegal connections to the overhead cables. Eventually, after two hours, we were free of the houses and driving across the pampa with the rising sun behind us. For the first few hundred kilometres it could be New South Wales, with fenced pasturelands, eucalyptus and paraiso trees and scattered farmhouses. But slowly the land toughened, the trees dwindled, and by afternoon we were driving through scrubby semi-desert in the rain-shadow of the mountains. Although the land was imperceptibly rising, it was not until early evening that the first low hills appeared: we were passing the southern end of the Sierra de San Luis. Eighty kilometres farther and the sun, still seemingly quite high, suddenly flickered out. It had set below the Andes, 160 kilometres ahead and invisible in the haze.

Our destination, the city of Mendoza, in the province of the same name, nestles against the foothills of the Andes. We arrived in the dark and left shortly after breakfast and I cannot remember anything about the hotel at all: I stayed in over twenty different ones during that six-week tour of Argentina's mining resources. Some were memorable: San Luis, for instance, where the hotel fronted the city's main square ('city' is a kindness) and the visit coincided with a triumph by the local football team. The supporters' cars were still hooting and careering crazily around the square

Ida who?

when dawn finally broke.

Jujuy (pronounced hoohooay) was another, if only because I was in a blur of exhaustion throughout my brief stay. I had arrived at 11 p.m. the previous evening, had left again in the small hours to be driven to a mine 160 kilometres away and 4,500 metres up in the Andes, made gasping ascents through the stopes and over the plant, and then back to Jujuy for a supper beginning, Argentinean style, at 11 p.m. My dinner companions had the benefit of a siesta and we lingered until 2 a.m. I left at 5 a.m. for the airport and had to be woken when the flight landed at Buenos Aires. Small change to the average traveller, but in a career made up of this it adds up.

From Mendoza next morning we drove up into the Andes, along a road that clung to the mountain sides and was so narrow that traffic was restricted to certain hours in each direction (at night it was two-way traffic: the appearance of distant headlights allowed drivers time to get into one of the few passing bays along its length). Our destination was a tiny underground bentonite mine, high up in the arid mountains. Bentonite is a clay with useful swelling and lubricating properties, making it valuable in drilling. The mine was a Zimbabwean-style smallworking, but in moist, slippery clay. The 'cage' for going underground in was a biggish wooden box, which was simply allowed to skid down an inclined shaft at the end of a rope, to the uneasy excitement of the novice passengers aboard.

The theory of bentonite creation is that it is arises from volcanic ash falling into salt water; the Andes were, and are, a very active formation. Tectonically, the mountains have been formed because the Pacific plate is diving down underneath them. This not only pushes them up, but the compression of the underlying mantle leads to magma being forced out through weaknesses to create volcanoes.

Almost all of these were, or are, on the Chilean side of the border, which from a mining point of view is unfortunate for Argentina. Chile has sixteen major copper mines on the volcanic rock known as porphyry. There is only one deposit similar to the Chilean copper mines in Argentina. This is Bajo la Alhumbrera, which was being prospected and drilled when I visited it. The

All poor together

distorting effects of national pride are vivid here; the mine is only about 200 kilometres from smelters on the Chilean side. However, to export the concentrates they are being sent by one of the world's longest slurry pipelines, which runs for 320 kilometres eastward, followed by a 800-kilometre rail haul to the Parana River and thence by sea to smelters in Brazil or Europe.

Bajo la Alhumbrera is, of course, far up in the bleak Andes, in a remote part of Catamarca province, more than a 160 kilometres from its namesake capital. The deposit was officially a new discovery, sold to the developers by the government. Yet it is marked by a prominent gossan, green with copper staining. A few kilometres away is a gold mine, Farallon Negro, which was discovered by the Spanish (or was it of Inca origin and did the Indians show them?) back in the seventeenth century. If they struggled up through these barren, dry mountains to find Farallon Negro, then they surely would have found the much more obvious Bajo la Alhumbrera. I am sure they did, but as it was lower grade than Spain's own copper mines at the time, they left it. This being the case it strengthens the view that, if a mineral deposit is visible on the ground, chances are it will have been seen and, at the least, scratched at by man many years before the official discoverers claim the glory.

The point that this discursion is weaving towards is that the upper Andes, such as the highlands around the bentonite mine or the bleak peaks about Bajo la Alhumbrera, are a wasteland of lava and scoria. Lower down there is a bit more vegetation, but the mountainsides there are composed of young conglomerates, mixtures of stones and mud, barely consolidated into rock.

Yet—and I had never thought of this until I actually encountered it—out of this nearly barren rubble heap comes the rich soil of the pampas, formed by the eroding rain. The run-off transports freshly mineralized silt from the Andes to the lowlands, giving rise to the agricultural prosperity of the whole of the eastern side of South America. The Andes, that immense, majestic formation, is a 6,000-kilometre-long pile of heaving volcanic detritus. It was an interesting reflection that there is no counterpart to it in Africa; only to the west and the north of the volcanic intrusions around the

Ida who?

Great Lakes are there some soils as deep and as good as Argentina's. The significance of this realization was not apparent to me at that time.

* * * *

The work in Argentina was in connection with a request for a loan for the development of the mining sector. Unlike most of the countries of sub-Saharan Africa, Argentina is a middle-income country, and so the work we were doing was for the World Bank proper. It does not seem to be widely known that eligibility for World Bank money is limited to governments assessed as being capable of repaying their loans. Such money is offered under conditions that are usually only modestly better than normal commercial lending rates, but the cachet of a World Bank loan is considerable, making other bankers ready to loosen their purse strings. However, those states whose per capita income falls below a certain figure—it was around $750 a year in the 1990s—qualify for soft loans from the International Development Association, which gives a convenient acronym.

'Soft' here means that IDA interest rates are way below inflation, perhaps 2% or 3% annually, and that repayments only start after a grace period of as many as twenty years. The World Bank operates the IDA system on behalf of governments, using the same specialists for assessing loan applications under both systems. However, while the World Bank raises its funds from bonds issued to the money markets, IDA gets its resources from governments—or, more correctly, taxpayers. So what we are looking at here is a dignified, defensible way of disbursing aid money on a large scale through the World Bank.

As of 1997, 79 countries with a combined population of around 3.3 billion were eligible to borrow in this fashion.[144] In its earlier years the Bank was prone to make something of those countries that had graduated from this bottom-of-the-financial-class situation, who were mainly amongst the younger Asian tigers. Since then less has been heard, partly because the numbers of graduates have been very small. More importantly though, to

All poor together

trumpet such successes might draw undue attention to the numerous drop-outs, whose increasing poverty has caused them to become ineligible for World Bank loans *per se* any more. These are principally in sub-Saharan Africa (average GNP per capita here fell from $660 in 1980 to $493 in 1996), and include many of the once-bright hopes, such as Zimbabwe, Zambia, Kenya, Ghana and the Cote d'Ivoire. Even some of the oil-rich countries, such as Nigeria, Cameroon and Angola, have joined them. It is easier, in fact, to list the countries that have not gone this way—Mauritius, Botswana, Gabon and so far, at least, South Africa.

The failure that this represents has been well concealed. Indeed, when Zimbabwe dropped out in the 1980s it was trumpeted there as a desirable development: now the country would get cheaper loans from the World Bank. In fact there would be no more loans from the World Bank; they would merely be administered by it.

While the same people assess the loan requests on behalf of both the World Bank and IDA funds, there is a certain difference in priorities. The money for the IDA fund has to be extracted from grudging governments every three years, and the experts tend to be both gimlet-eyed and fairly leisurely in their assessment of whether the 'loan' is needed or not. The common accusation that the Bank is guilty of foisting money on unwilling poor countries in order to make a profit is therefore unfair.

However, when the Bank operates as a bank, as in Argentina for example, there is a certain amount of pressure to get the money out of the door. By my IDA-formed standards Argentina's geology was not promising enough to make any special effort necessary to sell the country as a mining destination: better it concentrate on getting its agriculture competitive again. The loan went through, though.

One of Argentina's drawbacks from a mining point of view its is fragmentation into provinces, each with their own set of laws. Bajo la Alumbrera's billion-dollar investment nearly came unstuck on this point. Provincial officials at Catamarca decided to interpret the royalty basis on the total value of the output, despite it having been previously agreed with the federal government in Buenos

Aires that the net value would be used. This small alteration drove a great hole through the profits; Africa is not the only place with obdurate bureaucracies.

•   •   •   •

Nonetheless, dealings with bureaucracy in Africa in non-routine situations are always fraught. Even normal transactions involve great delay, incomprehension, general chaos. Southern Africa is usually not too bad (but try crossing the Limpopo Bridge between South Africa and Zimbabwe around Christmas), East Africa is rather worse, West Africa is bad, and francophone West Africa the worst.

So I invariably get a sinking feeling when arriving in Abidjan and entering that character-building institution, *l'aéroport Felix Houphouet-Boigny*. Twice previously circumstances had led me to arrive here without a flight booking or a visa, and on both occasions I had lost out, once badly. The first was tolerable: the flight I came in on from Mali was too late to connect and I had to wait three days in the country to get out on the next one going in my direction (on that occasion the visa problem was overcome by a Air Ivoire official who slipped me in through an officially closed door, probably to avoid the problems that I looked like creating). That only cost me three days' hotel bills and a surcharge on the ticket, which my client (a World Bank trust fund, as it happened) said it had no money to pay me with.

The second was more serious, and arose from concern for my welfare from a Ghanaian client. Most unwisely I was contemplating starting from West Africa with a heroic itinerary and tight deadlines. I was to fly to Togo from Accra on Ghana Airways and board Ethiopian Airlines there, which was to get me to Zambia the following evening. That night in Lusaka I was to collect a motor car and next morning drive 350 km southwards to Livingstone. There I would turn west and drive 150 km on a little-used and badly rutted road that paralleled the Zambezi, use a ferry to cross the river into Namibia at Katima Mulilo and then drive 1,400 km through the Caprivi Strip and down to Windhoek. There, three

All poor together

days after leaving Ghana, I was to rendezvous with (and deliver the car to) the rest of a Zambian-based team. They had assembled there to start travelling to mines on a project that was looking at what mining was doing to the environment in that country (not very much, really, but then again I am, like most miners, unrealistically defensive about what we inflict on it).

Togo had been on the point of boiling over and the army was on the streets, or so it was said. My client thought that I should not travel that route, better to go through Abidjan (coincidentally while writing this on Christmas Eve 1999 there was a military coup in Cote d'Ivoire). Luckily, he said, he had a sister who was a travel agent. But the endorsement she made was, it turned out, not accompanied by any re-booking or validation and so was not worth the paper it was written on. For something over a thousand dollars—my entire stock of traveller's cheques—I had to buy a whole new ticket at the airport. I didn't get repaid for that either.

Now, for the third time, I was arriving in Abidjan with a ticket for a flight that either did not exist or had already left and with no visa, but now at least, by God, I had a credit card. I was almost sanguine.

I was in transit, coming in on an international flight, and needed to get to the transit desk in the international departure hall to start the gruelling process of getting on another flight. But, as I suspected, it was not just a matter of entering at the transit door. I slunk off in that direction, but was smartly told to return and join the straggling queue at the immigration desk.

I protested to an official (a porter, actually) who said that it was all right, all I had to do was to put $10 in my passport and give it to the immigration officer and he would stamp it. That was all. Indeed, just give the porter $20 with the passport and he would get it done for me.

For reasons connected with difficulties in changing traveller's cheques in the remoter parts of Burkina Faso I had only two or three dollars in cash on me. But I did have plenty of time, two, perhaps three days even. So I gave the porter a brisk lecture in the evils of corruption, or at least as brisk as my French would allow. He shrugged his shoulders, bemused, but did not abandon me.

Ida who?

Eventually I got to the official in the grubby glass booth and pushed my passport over.

'Je suis de passage, en transit,' I said.

Unsmiling flicking through of pages. 'Visa?'

'Pas de visa. J'ai arrivée de Ouaga. Je depart pour Johannesburg demain. J'attendre dans l'aéroport.'

He paged through the passport again.

'Dix dollars,' muttered the porter in my ear.

There was a long pause. The official flicked through the pages ever more slowly, studying the complex, overlapping mixture of visas and stamps. Behind me the queue was getting restive.

'Billet?' I handed over my ticket. He stared at it.

'Attendez la bas.' He pointed to a spot beside the booth, and so I waited while the rest of the flight was dealt with.

Eventually they had all trickled through. I continued to wait. Then I was summoned and waved through to outside. The official— a smartish young man in khaki with silver stars on blue epaulettes— left his box to come after me, pointed at my passport safely in his hand and told me that I must have a visa. I protested that I was in transit, it was not necessary. I did not wish to enter Cote d'Ivoire.

But, as he pointed out to me, you are now in Cote d'Ivoire. Without a visa he must take me to the police. The argument became very boring and repetitive, enlivened only by his suggestion that for a small fee he could arrange things. I remounted my high horse, and in the face of this scolding he left with my passport, giving instructions to wait. I took out a book and began to read. This encounter was going to take time, perhaps all the rest of his shift.

I was in the entrance to the domestic part of the terminal, congested with people. After an endless half-hour my captor returned with, surprisingly, a chair. So I sat, as imperturbably as possible, amongst the curious throng. It was, I suppose, only another hour or so before the official returned, although by that time I was starting to find my book rather difficult to concentrate on. In aid-speak, we revisited our differences.

Then it was over. A shrug, and he gestured at me with my passport to follow. We marched smartly over to the international

All poor together

entrance, past the crowded check-in desks, the benches of the *douaniers*, the immigration booths, the X-ray machines, the duty-free area and into the departure hall. He returned my passport. I controlled my relief with difficulty and ostentatiously tipped him two dollars.

'De rien,' he said, but swiftly pocketed the note, and we parted smiling, he thinking that I was an idiot and I thinking that he was corrupt. I turned to embark on the next struggle, this time to get my ticket endorsed across to another carrier. But that, you will thank God, is another story. Enough to say that there was plenty of time to think.

• • • •

The pieces of the jigsaw that might solve the problem of, first, why sub-Saharan Africa is so under-developed and, second, why it does not respond to development assistance, are almost all assembled. Here they are as I stumbled across them (and I am sorry to be didactic, but they do lend themselves to bullets):
Starting with Africa—
- The correlation between education and growth is missing: against intuition and all the evidence from elsewhere, increased schooling was associated with lower, not higher, personal incomes
- For mankind, it is the most disease-prone area on earth.
- Yet, certainly since the arrival of Western medicine, it has been in the grip of a population explosion that has caused, for example, Zimbabwe's numbers to grow more than twenty-fold in a century.
- Its most fearsome disease is a sexually transmitted one, AIDS, whose spread has defied common sense, to the point where it will stop the population explosion at   enormous human cost.
- The culture across the continent is the same—the bride, not to mince words, as a baby-generating property, bought for that purpose, often inheritable within the   family and returnable if not functioning as such.
- The family itself is a much bigger social unit than in developed

countries, with the ties of polygamous marriage bringing loyalties and obligations to a host of remote relatives.
- Geologically, Africa shares with Australia the distinction of being the least tectonically active of continents.
- So its soils are not replenished by mountains a-building, resulting in their being leached and infertile, often capped with a hard laterite seal.

And now, what was going wrong with development assistance?
- At the lowest level, aid does not work because very often the skills required, if not unique to Africa, need a special slant for them to be transferred properly. My knowledge, used to keep Buckreef staggering futilely on, happened to fit; it was specific and practical. Most aid-workers are not so equipped.
- More generally (and, again, as Buckreef showed), the environment in which aid exists—if not donor government to host government, then at least donor government to a body sanctioned by the host government—stops it from functioning. The donors have their politicians, their taxpayers and their reputations to look to, counterbalanced by the need to spend this year's budget. There is only one item on the host government agenda: to stay in power and access its benefits. Aid provides political largesse that can be used to buttress this, if only in the form of a few jobs to hand out.
- Beyond this again was a feeling—nothing quantifiable—that aid did not 'take' as it should in Africa. When a project finished, it was usual for its momentum to be lost. Projects and institutions set up in a white heat of enthusiasm became bywords for lethargy and inefficiency. Without the donor representatives being involved, there was an acute case of something that resembled the 'not invented here' syndrome. Perhaps the trouble in Africa was that nothing very much had been invented here.

My ruminations on this gloomy picture in the departure lounge of the airport tended to be interrupted by people running their baggage wheels over my toes and the need to concentrate on the

All poor together

echoing, incoherent loudspeaker announcements. But after having erratically worked my way through the pieces (not that they were in anything like the tidy array above) I concluded that it might be best, to start with anyway, to assume that the problem lay with Africa, not aid. After all many other parts of the world had used development assistance successfully to get up to speed; South Korea is a good example.

That decided, then the ubiquity of this failure on the continent suggested that the common denominators must be important. The most fundamental of these would be geology: Africa the quiet continent. Then culture. People live in their culture, and I had found that Africa's is essentially the same all over. This raised the wife-as-property feature again.

At about this time a large Lebanese family set up a small laager of suitcases opposite me. The mother and older daughter (or at least I assume that is who they were) had gold bangles cluttering their arms, and I remembered that in Nyerere's Tanzania this was said to be a way of getting dowry money out of the country to the husband-to-be in Bombay or Calcutta.

Now that was a funny thing about Africa: in most countries where there are financial transactions involved in acquiring a wife, it is more a question of the wife bringing a dowry along with her, to add to her husband's assets. Yet in Africa (and was it only Africa?) the aspiring husband had to buy her from her father. Why so? What was so fundamentally different about India, for example?

Now I knew the answer to that. The Himalayas, tectonically a continental impact feature, provided rich soils around the river systems of the sub-continent. These soils had supported the green revolution in rice and wheat there. Africa's staples are maize and cassava; rice comes a poor third, wheat almost nowhere. Yet the wherewithal for its green revolution has long been in place: Rhodesia's national average yield of maize increased from 1,146 kg/ha in 1946–50 to 4,725 kg/ha in 1976–80 thanks to the development there of hybrid varieties there, coupled with improved fertilizer use.[145] But the revolution had faltered on the barren soils of the continent and the average peasant's inability to afford hybrid seeds and fertilizer .

Ida who?

Historically, apart from animal dung, there never was much fertilizer, anyway. When Africans began to grow things to a purpose, they usually were forced to adopt a slash-and-burn technique, which enriched the soil with the minerals from the ashes of the indigenous vegetation. But for most of mankind's history—African history—mankind was a hunter-gatherer, living a drifting existence as he followed the wildlife. The Khoisan of Southern Africa are the living inheritors of that long tradition. Yet whatever you did to survive in Africa, you could not do it in the same place for very long. Not unless you were mining, or trading gold, or some such. Then you settled down and built permanent lodgings of exfoliated stone in Zimbabwe, of hardwoods in Ghana or sun-burnt bricks in the Sahel.

My thoughts were diverted back to the crowded departure hall about me. Through it strode what the people of Zimbabwe would call a *chef*. Not a tribal chief or a cook, but a Big Man; the name is Portuguese and originated in Mozambique amongst the exiles there during our war. Once a term of respect, it is now commonly used in a disparaging sense of one whose political and business clout alone has enabled him to get rich. This chef was evidently flying in a private jet, and he led a little party of bag-carriers and gofers to one of the departure doors, preceded by an official together with another gofer clutching a handful of passports. Many of my clients are *chefs*. This one was not fully developed; he was not particulary fat and was not talking emphatically into a cellular phone. However, he was wearing dark glasses, was excessively nattily suited for the occasion, had a number of gold rings on his fingers and amongst his entourage was an extremely smart and haughty lady who was surely too young to be his wife, and too glamorous to be his secretary. There was a little wistful silence as the group sped through, a sort of collective sigh. How nice, was the common thought, to be *him*. Or *her*.

It was then that the penny dropped. Man, the status-seeker. If you are going to have to keep moving around, then the assets that prop up your image must be able to move as well. Which, in a society without motor cars, private jets, or even horses, limits them to three. Livestock. Wives. Children.

All poor together

. . . .

After that it was easy. If children are status symbols, then more is better (I have never completely bought the theory that Africans need lots of children to support them in their old age; why more there than other human societies require? What was sure was that the high level of endemic diseases had led to babies being terribly vulnerable, requiring perhaps twice as many to be produced compared with elsewhere if the family was to survive). Then, if a father owns the mother, so much more so does he own the children.

To have lots of children requires lots of mothers, hence polygamy. Polygamy comes with, literally, a price. Parents who have themselves invested time and effort in bringing up daughters are not just going to give them away in an environment where she is first and foremost a baby-machine, who can be corralled in with other baby machines to generate wealth and status for the common husband. Mothers, or rather potential mothers have a value, and have to be paid for. The only asset available as currency for this transaction is livestock.

Cattle are therefore a money and they obey the two laws that control individuals' reactions to it; (1) to have more is better and (2)—the one noted by Sir Thos. Gresham—hang on to the good and get rid of the bad. If a potential wife is valued by her father at forty head, you don't send across your fattest and sleekest. In the matter of bride price it is quantity not quality, the reason behind the despair of aid workers in animal husbandry, who thought they were working in an environment where people wanted to breed livestock for food or for sale.

So African culture was, as might be expected, the outcome of an adaptation to Africa's physical environment. Having pieced this together, I was feeling quite cheerful when, late at night, my flight eventually was announced, and I left behind (but not forever, unfortunately), the crowds, the noise, the congestion, the tensions of the *Aeroport Felix Houphouet-Boigny*. Third-world travellers will agree with me, I think, that the happiest words to hear are 'Doors to automatic and cross-check'. Now at last you can fasten your safety belt, with a goodish expectation of getting to where you want to go.

Having spent too much time in the past couple of days doing nothing very much except low-level worrying, I found it difficult to doze off that night. Besides, there was a nagging feeling that I had missed something. I had a neat theory based on geology about how and why African cultures were so homogeneous but, apart from the business about animal husbandry, I couldn't see why it should make aid impossible to implement successfully. There was a hole, a big one, in the middle of the jigsaw.

To go back to the problems with aid, I had to agree with the received view that it didn't work because it was a matter of poor 'governance', that po-faced word. What was there in this culture that caused governments to have such an arrogant attitude towards 'civil society'? The only occasions that any restraint is shown is prior to elections. Then all the old political hacks make tentative gestures towards the *povo*, mouthing phrases about achieving economic independence, reducing taxes on the poor, soaking the rich. It was a sobering thought that almost all democratically elected governments in Africa (not many) had sustained their hold on power by promising the redistribution of wealth. It is a position that almost guarantees re-election when the key electorate is the communal-living rural poor; there are no votes to be gained by those to the right of it and anyone who manages to squeeze in to the left can be safely locked up as a dangerous anarchist.

I sought mentally for my list again. First, to dispose of what economists would describe as the exogenous factors. This is the provision of inappropriate or unsuitable skills and equipment. My false success at Buckreef had nothing to do with the Tanzanians, everything to do with Sweden being the wrong place to source gold-mining expertise. It is to their considerable credit that when the crunch came, they didn't blunder on, as often happens, but recruited me.

This leaves two possibilities. First, the environment in which aid operates. This is irretrievably warped by donor government and host government policies and politics. Buckreef should not have been funded as an aid project: if the private sector was not

All poor together

interested in it, then nothing further should have been done. But once a host or donor government backs a project, good or bad, withdrawing requires a level of political courage that is usually lacking on both sides. Aid-supported institutions in Africa never close down; they linger on indefinitely, no matter how useless or irrelevant they have become.

But these are still, partly at least, exogenous reasons. Even if it is a worthwhile project and the skills or equipment to be transferred are appropriate and are successfully delivered, failure is the usual outcome. As numerous people have said to me in one way or another, 'once the expats go the whole project will fall apart'.

The problem, then, is due to the third aspect; an endogenous factor or factors. Bluntly, something in the African outlook conspires to thwart the effective implementation of development assistance in Africa. It is likely that it is something to do with its polygamous, semi-nomadic, extended-family culture.

Clearly this culture does not sit well with the more impartial, business-orientated traditions of developed countries. More than anywhere else, in Africa it is who you know, not what you know, that counts. The extended family is the main loyalty unit. The tribe (in the sense of a group with a common language or dialect) is next. The state—almost invariably an imposed, artificial construct in Africa—comes last by a long way. Somebody outside the family or (worse) the tribe is a person of no account, if not an actual enemy. And aid workers from outside Africa? Blinkered, affable aliens, who understand nothing except how to manipulate their marvellous machines, and who always fly away again after a couple of years back to their own fabulous homes. They have no reality, no relationship to life as lived in the African world. In any time scale apart from the short term, they can be disregarded.

And there is the answer. Aid agencies deal with governments in Africa, under the impression that they represent the people of the country. They do not; they represent themselves, their families, their tribes. Only for the sake of appearances do they assume the dignity of national leadership.

It was not the whole story, but now sleep finally arrived.

Ida who?

• • • •

The credits for the in-flight movie were rolling silently up the screen when I awoke. Then it reverted to the earlier map of Africa, where the red line of the flight path, last seen somewhere over the Gulf of Guinea, was still creeping southwards. Now it was near the Victoria Falls. It was 5 a.m. outside and I reset my watch. On easing up the window blind a great cloudscape was revealed. Below and westward a flat red-tinged deep-pile carpet stretched away unendingly, through which erupted a dozen billowing thunderheads. In their shadowed bases lightning flickered, but the peaks were in full sunlight, shining gold. Home was down there.

Chapter Twenty-one

A THOUGHT AT SUNDOWN

'I've never been poor, only broke. Being poor is a frame of mind. Being broke is only a temporary situation.'
—Mike Todd.

The chief had undergone an operation for cancer in his larynx, and spoke faintly and huskily of his gratitude through a throat microphone. On the wall of his wooden shack was an elaborate certificate of commendation from the late President Tubman of next-door Liberia. The chief's gift, a bottle of gin discreetly wrapped in the *Freetown Democrat*, was handed to an aide and we got down to business.

This was a job with no less than three clients, all sitting with me on a narrow wooden bench, on a sweltering October day. They were Cato, a Norwegian, Stephen, an up-country Sierra Leonean, and a businessman from Freetown. The latter was descended from freed slaves, brought across from Jamaica in the nineteenth century. His name was Joe Cohen.

In theory this trio had a concession from the government that would allow them to prospect for gold in the area. In practice we could only do so if the chief agreed, hence the meeting. The language used was Krio, a Creole lingua franca that had strayed so far from its English roots that only by concentrating very hard could I get the gist of what was going on.

After about an hour, the main details were agreed. A cash payment of 5,000 leones (about $10 then) up front and thereafter 5,000 a day for every day that prospecting was going on. So, said the chief, please give my assistant here 10,000 leones. We did not ask for a receipt; our unimpeded presence on the chief's land was its own evidence of payment. A summons went out and a young man appeared; he was one of the chief's sons and his job was to let us get on with our work and, of course, to tell the chief in

A thought at sundown

advance if we found anything.

The results were disappointing. There was said to be much alluvial gold in the area, somewhere north-east of Bo. But only women were working the streams, always a bad sign, and my own pannings showed only a few colours, no tail. Nor were there any resident buyers; the gold was being taken to Bo for sale. We needed to dig to bedrock to see if anything was down there, but this, it seemed, was not part of our agreement with the chief. There was, however, a pit nearby where 'much gold' had been extracted, although unfortunately it was now full of water. But if we bought a pump from the chief, who had one handy, we could check for ourselves.

I decided to call this bluff. I stripped off and dived down to the bottom of the grey pool, surfacing with my hands full of sand. When I panned this there was nothing, no colours at all. The chief's son said that there was more all around; perhaps that particular hole had been worked out. Later we trekked up into the hills, looking for the reefs that the gold had come from. They were there, but thin and scattered. Worse, even in this desperate land nobody was working them, just a few abandoned scratchings.

We had placed an order for palm wine at a grog shop in a village on the way up into the hills, and drank it gratefully on the way back that evening. Palm wine, like retsina, millet beer and South American maté, does not travel well but is great on a very hot day. I told Joe that there was probably no gold of the sort that would make his fortune in the area.

'So why the women they panning, then?' he asked, reasonably.

'Because they can survive on a hundred leones a day. Besides, they have children to feed.'

The country was as poor as any I had seen. It was late summer and the rains were coming to an end, but the maize was yellow and spindly. The clothes of the people were falling apart, there were no bicycles and swarms of flies buzzed on every child. A thought struck me.

'The chief controls the land, doesn't he?'

'Oh yes, he say who farm where.'

'The people would be a lot better off if they owned their land.

All poor together

They could raise money for fertilizer every season. Could the government do that?'

Joe snorted. 'These people would sell quick quick and piss the money against the wall.' He was a capitalist from the coast, one of the then ruling class. Yet Stephen, who was, after all, from these parts, yet evidently had the rare ability to pull himself out of destitution into modest affluence, did not disagree.

That was my last trip to Sierra Leone. Even then, in 1992, the diamond fields of Kono, not far to the east of us, were under threat by rebels based, it was said, in Liberia. Since then, as in Liberia itself, the ethnic gap between the commercially adept, property-owning elite, descendants of freed slaves, and the indigenous Africans has degenerated into a civil war of appalling viciousness.

Ethnicity creates wealth but drives conflict, and not only in Africa. An insight into the forces behind East Asia's success came to me on the schooner wharf in Jakarta. Long ago I had travelled on such a Javanese-registered schooner overnight from the dhow wharf at Dar-es-Salaam to Zanzibar. Like all of them its sails were used only intermittently, for it sported a big rumbling diesel engine, which we needed to butt through wild seas on that trip. It was a journey memorable only for its seasickness.

These schooners had made a longer voyage and were deeply laden with hardwood timber from Kalimantan. The sun was glaring down on the keyside as sweating Amboinese labourers trotted down in pairs, laden with the heavy planks. They were supervised by a foreman, a Javanese, who arrived on a small motorbike. Later he was joined by his boss, who arrived in a smart new Toyota, and was an exponent of the theory of management by nagging. The boss was Chinese, all the rest were indigenous Indonesians. When Suharto was deposed two years later, it was the Chinese who, in a long-recurring practice, got it in the neck.

Apart from Japan and South Korea, almost the only thing the Dragons have in common has been the entrepreneurial acumen of the Chinese. There were, I found, worrying parallels between them and the Indians and Pakistanis of East Africa, the Lebanese of West Africa and ourselves in the south.

Yet, time and again, one finds exceptions to racial stereotypes,

A thought at sundown

Indonesians or Africans with an entrepreneurial ability, who had used it to become relatively wealthy. Sometimes they were rich enough to dabble in gold mining, and to use people like me for advice. Apart from that flair for business, did they have anything in common?

Well, they operated in terms of money rather than relationships. They were not necessarily beholden to political influence for their wealth; come to think of it most of them were profoundly distrustful of their governments. And they were property owners, there were no exceptions to that. I had stayed in their, often lavish, houses.

Then, finally, revelation came. Of course, this was the missing piece of the jigsaw, the gap that had worried me in the night flight from Abidjan. Most Africans may be living off the land but it has no value for them; the chief has simply allocated it. Yet wherever societies have prospered without some special resource advantage, such as oil, individual wealth had been created by the concept of land as private, tradable, property.

Mining pointed the route. If, in Zimbabwe, it had not been possible to have mineral rights as private, tradable property, then the country's mining industry would be just like elsewhere in Africa to the north of us. We would have a smattering of very large mines, whose mineral rights had been painfully, individually, negotiated by their foreign owners, and a multitude of illegal panners and diggers, with nothing in between. Instead we had more mines—real mines—than the rest of the continent put together, hundreds of them.

Thinking about it, in an economic vacuum like Africa was there any way other than through private property ownership by the poor, that poverty could be alleviated? In most of the continent the very poor are on communally owned land, and giving title to the inhabitants would, or rather should, preferentially increase the wealth of the poorest. Armed with this security and a widespread banking system to give a multiplier effect, money would be created.

The basic resistance to land titles—and one which is heard more from aid donors than from the Joe Cohens of Africa—is the certainty that it will lead to many owners selling theirs to shrewder

All poor together

or wealthier individuals. There are two ripostes to this. One is that this is exactly the revolution than has propelled countries into prosperity. After delays and unpleasantness (which industrial-revolution-type problems surely—surely—the economists have cracked by this stage), the wealth that has been created by giving a value to fixed property would emerge as funding for enterprises and markets in the villages, towns and cities.

The second counter-argument is that the drift to the towns is already in full swing anyway. City growth in Africa is the fastest in the world. At around 5% annually it is rising faster than the population as poverty in the rural areas deepens. Yet, unlike elsewhere in the world, this, as the World Bank has noted, is urbanization without economic growth.[146] The majority of the incomers are squatters; they have almost no possessions beyond what they stand up in and almost no hope of work in the formal sector. With so little money about, the markets are tiny, and without widespread wealth there is negligible competition for the few capitalists around. They have usually worked through political (for which read familial, tribal) connections to secure a monopoly.

In any event, landlessness has not meant joblessness elsewhere, so why the concern about introducing fixed property rights to Africa?

The reason appears to be that it will destroy a cultural system that may be irrelevant to the point of suicidal, but which carries with it the comforting concepts of the extended family and of communal and tribal solidarity. Not incidentally this makes for an inward-looking, subservient population, convenient for exploitation by its politicians, who have been largely unconcerned about the value of private property to the people. Apart from themselves, that is.

Yet the replacement for this culture is materialistic and individualistic and deeply repugnant to most Africans' outlook. Even Westerners call it the rat race. Is there a less Thatcherite route out? Surely, given enough inputs—fertilizers, irrigation pumps and so on—the land can be made fruitful again. After all India feeds a much larger population on a fraction of Africa's land surface.

Certainly, and India is a sub-continent of, largely, private land

A thought at sundown

ownership; there is little need to find substitute wealth symbols like wives or children. Why, parents there pay husbands a dowry to take their daughters off their hands. Demanding a bride price would seem grotesque.

A grim choice now faces the development assistance business. Behind it stretches decades of wasted interventions. Ahead is a vista of dwindling support as donor (meaning voter) disillusionment increases. And all the time in Africa the relentless, unsustainable growth in demand for food, for jobs, for schools, above all for hospitals and the treatment of AIDS, will continue. At the present pace of deterioration Africans will be as poor as they were in Livingstone's time in about the year 2040.

Thus Africa's problems. For the aid industry *its* problem is that very few people in Africa are wholeheartedly in favour of private land ownership; the peasants because they are properly suspicious of anything that will demolish an ancient heritage, with its comforting tradition of extended family responsibilities, and the chiefs and governments because it removes their social and political clout.

One result is that in Zimbabwe freehold commercial farms are being purchased by the government to resettle landless peasants—but they are not being given transferable title and so remain, effectively, without assets. The political pressure for forced commercial farmland acquisition in Zimbabwe is understandable, as whites own about 70% of such farms, giving a severe racial imbalance in land holding. However, by failing to give the new, black, occupiers ownership the government is condemning them, along with their (necessary, as I now understood) prodigious number of descendants, to rural destitution.

•   •   •   •

Of course, my revelation was somebody else's accepted wisdom. Quite a lot of people, in fact. In 1999 the World Bank published a review of land policy[147] that summed up what happened when what I thought should be done was done. It has over a hundred references. Curiously though, its abstract said that it had been recognized that 'communal tenure systems can be more effective than formal title'.

That turned out to be a gross simplification. Under certain circumstances—where there is a low population density, where only arable land is involved (as opposed to common grazing or forest or fishing grounds), where inheritance is accepted and where cash land transactions within the community are allowed—communal ownership seems to work pretty well. This is quite a string of qualifications; I don't know of anywhere in Africa where they all apply. The paper records that in Zambia, with one of the lowest population densities in Africa and some of its better land, almost half the communal farmers there feel that their land tenure is insecure and would be prepared to pay something for formal title.

Besides, simple population density is misleading. When the poor quality of the soil is taken into account, many of the vast, empty countries I had worked in turned out to be relatively densely populated.[148] On this basis Mali and Burkina Faso are twice as densely settled as Indonesia, Niger more densely populated than India or Bangladesh and Kenya has one of the highest agricultural population densities in the world. That this will lead to disaster had not gone completely unnoticed:

> These people face a critical dilemma: a central element of their traditional farming system—the ability to shift around on the land—is being eliminated by population pressure, yet they continue to use the other elements of their customary production systems.[149]

A vivid piece of visual testimony to this comes from a satellite photograph of Zimbabwe at the end of the dry season. It shows a patchwork according to land titling. The communal lands are pale, over-cropped, over-grazed areas, the commercial farmlands (mainly white-owned) are still thick with vegetation. What is significant though is a third category. These are the 'Purchase Lands' where communal peasants with some farming ability were given title during the colonial period. Like the commercial farms, they stay green in the annual drought.

A thought at sundown

The weight of evidence in the World Bank paper on land policy for the opposing argument—that the absence of secure and transferable private land title is economically crippling—is immense. Mexico, China, Russia and India are quoted as well as Africa. The latter has got some success stories to show, such as the redistribution of the 'white highlands' in Kenya, but these are limited. Elsewhere in Kenya, for instance, tribal tradition, as embodied in the continuing influence of chiefs and elders, has largely thwarted the creation of land markets.[150]* It is the countries that have successfully entrenched their private fixed property rights that have boomed. Turkey doubled its average income in twenty years (1957–77), Brazil in eighteen years (1961–79) and South Korea in eleven years (1966–77).[151]

The most dramatic example is China, which took only ten years to do this (1977–87). This is the real Chinese revolution, not the slaughter and tyranny of the previous ones. It began in the late 1970s, when 15-year tradable lease rights were given to the peasants for their land, along with the ability to sell their own output. So successful was this step that it has now been replaced with 30-year contracts. Everything in modern China's success flowed from this simple, limited formal tenure. By contrast Russia, whose communist apparatchiks have subverted land reform, has seen its economy crumble.

So what the record does demonstrate, unequivocally, is that security of tenure and the tradability of that tenure are fundamental to national success. As for the pissing against the wall hazard, the authors of the World Bank paper on land policy make the following sober, if prolix, judgment from their review of the evidence:

> Precluding beneficiaries of land reform from renting out or selling their land is likely to prevent adjustments that reflect the settlers' abilities, and could, if combined with restriction on rentals, cause large tracts of land to be

---

* The Masai Moran with whom we had the fraught confrontation at Lolgorien in Kenya in Chapter 9 were fierce enough in defending their own tribal interests, but at the same time were enthusiastic buyers of land recently made available for freehold purchase adjoining their area.

All poor together

underutilized. The goal of preventing small landowners from selling out in response to temporary shocks would be better served by ensuring that they have access to output and credit markets and technical assistance and by providing them with safety nets during disasters to avoid distress sales.

There is a further factor that was not covered in the paper. This is that for Africa to become a food-exporting continent again, it must have fertilizer. As noted, for tectonic reasons the soils are too leached to sustain production. In the *Herald Tribune* of 19 August 1993 an agriculturist, Dr Richard Critchfield, wrote an article entitled 'In Africa, avoidable disaster'. He argued that fertilizer subsidies, swept away as part of the World Bank/IMF dislike of all such arrangements, must be reinstalled in Africa if the continent is to reverse the agricultural crisis there.

The agricultural contrast between Africa and other parts of the world is certainly dramatic. To fly from, say, Denver to Atlanta, is to fly for two thousand kilometres (1250 miles) over the great checkerboard of America's breadbasket. In the past decades the analogy has become even more accurate, the proliferation of centre pivot irrigation systems has created a vast stretch that is marked with green circular counters.

Now fly over Africa for the same distance—say from Nairobi in Kenya to Harare in Zimbabwe. The landscape below appears empty apart from a few straggling red dirt roads and a huddle of huts or corrugated iron roofs surrounded by endless bush. The farms are usually small areas of cleared land around a village, flanked by older fields with meagre regrowth. Only for the last 160 kilometres (100 miles) as you approach Harare do farmlands appear that begin to resemble America's abundance—and consume fertilizer at the same rate.

The vacuum over which you travel is typical of the continent; it is possible, for instance, to fly westward from Addis Ababa in Ethiopia to Dakar in Senegal—6,000 kilometres (about 3,750 miles)—without seeing more than a score of farms or plantations as an American would understand the term.

We have seen the disarray that the failure of their economic policies in Africa has left the World Bank and the IMF

A thought at sundown

policymakers. As a result of this there has been some softening of the stance on fertilizer subsidies, particularly as Africa's agricultural output relative to its population has not improved. But their instinct is correct: Africa should be able to profit from agriculture without subsidies. Yet so is Dr Critchfield's: without fertilizers there will be no African agriculture above subsistence level.

To reverse the trend, it is not necessary to fret about fertilizers. Rather give the peasants transferable title to their land, create a land bank to turn that potential wealth into loans for agricultural inputs. The repayments can be used to fund projects to employ the surplus people no longer needed when subsistence farming turns into the commercial sort. Sustainable agriculture will follow.

Something else will, as well. When secure fixed assets are available, the incentive to create mobile ones is reduced. The new men of property (there will not be many women to start with) will face a choice between the generous and prolific African family of tradition and a new selfish bourgeois culture with its sparser, wealthier offspring. If they are like my clients, the population explosion, already dampened by AIDS, will slowly vanish. The demographic transition, so confidently expected in Africa and so long overdue, will finally arrive.

However, private land ownership will not in itself be enough to halt quickly the now-catastrophic cultural imperative for large families. Kenya has thriving, if overpopulated, private smallholdings to the north and west of Nairobi, and only now, thirty years after they were created, are the families living on them starting to reduce their birth rate.[152] There must be a catalyst to accelerate this cultural change, and logically it has be in the form of making loan finance readily available at the village level to the new landowners. This facility would bring home the startling new idea of wealth in terms of fixed property.

There is a precedent. The opening up of the American near-West in the early nineteenth century was achieved through finance from banks that sprang up at every crossroads (or so it seemed at the time). Certainly many of these banks failed or vanished, complete with their depositors' funds, but usually not before the more enterprising had used them to acquire land and create economic enterprises. The descendants of those banks that succeeded still exist, often as money-centre institutions in the major cities. The growth of America has always been spurred, not hampered, by its plethora of vigorous small local banks.

African versions of this phenomenon need to be created to accelerate the rate of cultural change. Banking in Africa is currently the preserve of heavily regulated monoliths with a scattering of branches. The real need is for very many small village savings banks to be licensed to members of the continent's entrepreneurial class, replacing or transforming the moneylenders in the villages so that the newcomers can take deposits, as well as lending money at less than usurious interest rates.

It still will not be easy, however. Interest rate reduction is another part of the changes that are urgently needed if Africa's land tenure revolution is to work. At the present time on the continent interest rates are commonly high enough to daunt entry into the economic sector by any enterprise that is not a monopoly. Rates of below 20% are rare; many range from 30% to 150% and above. These interest rates, coupled with highly centralized banking structures, are used to suck up what spare cash there is from the private sector and deposit it into government stocks, whence it can be used to pay the ministers, the civil servants, the army, etc.

The villain of this gluttony for cash by African administrations is another cultural imperative. The semi-nomadic culture enjoins intense family and tribal ties. As a government or parastatal official you are defying your deepest instincts if you fail to honour this commitment. There is a valuable commentary on this from Dr Nkosana Moyo, who was quoted on the role model question at the head of the last chapter. In the same paper he said—

> I want to touch on the issue of aid fixation. In African culture there is the family-based social security system. Within the extended family we traditionally look after the less fortunate members of the family. Both the giver and the recipient of this assistance are brought up to take it as given.
> Could it be that this erstwhile useful social security system has predisposed us to begging?[153]

A thought at sundown

Certainly these commitments meet head-on with the IMF insistence on sound money, open government and positive real interest rates. Because international banks will not go where the IMF fears to tread, most sub-Saharan states cannot raise any more loans anywhere before meeting this condition. Yet without literally putting themselves personally at risk, Africa's leaders cannot default on their tribal and political obligations to the horde of unproductive functionaries and soldiers to whom they are culturally committed. So, by hook or by crook, the latter continue to get paid, money is printed, inflation rages and interest rates soar out of sight. It is a development that cripples the productive sector and accounts for the stagnation of output in countries whose private businesses and farms have achieved some modest economic success, such as Cote d'Ivoire, Ghana, Zambia, Zimbabwe and Kenya.

Another objection to introducing private land tenure is that the technical problems associated with surveying and registering innumerable small plots on a continental scale are insuperable. This cannot be right; the tools for the data base management and rapid surveying required for mass land titling are now at hand—computers and centimetric global satellite positioning (GSP). This is just the sort of thing development assistance should be able to do very well.

One last example from mining to prove it. Zimbabwe has a system of registering mining claims—full-blown transferable mineral property rights—that requires only a 1/50,000 map blown up to 1/25,000, the universal grid system and an inspectorate to check on the location of beacons. Because of this simplicity there are over 25,000 blocks of claims in Zimbabwe, and the system complements the mining law by allowing the rapid acquisition of tradable fixed property rights in the event of a mineral discovery. It does not require computers or satellite positioning units; it has been in use for nearly a hundred years.

• • • •

The model of Africa's poverty trap and the way out of it appears

All poor together

complete. At its heart is the absence of any tradition of fixed property as assets. I now believe this must be because Africa's sedate tectonics has led to a vast, flat stable continent of well-leached soils, bereft of the mineral replenishment that, for example, the Andes and the Rockies have given to the Americas, and the Himalayas to Asia. Apart from areas of moderate igneous activity about a relatively short axis between Kenya and Zaire, survival in this environment has always depended upon a semi-nomadic pastoral or slash-and-burn agricultural culture, where all assets have to be movable. This limits them to livestock, wives and children, hence overstocking, polygamy and, once a modest amount of modern medical attention was available, the exponential growth of the population.

A consequence of this is that only a few countries (for example Kenya, Zimbabwe, Nigeria, Ghana and Cote d'Ivoire) have even a moderate acceptance of private, tradable, fixed property. Most of Africa outside of its capital cities is still permeated with passive communal ownership, which sits, to use another geological term, unconformably with Western economic schemes, with their core assumption of the force of individual self-interest. It is as distressing for me to say this as it will be for you to read it, but African development will only get under way when Africans are drawn into the rat race, when they become Forsytes, basing their wealth on their ownership of the land they once merely grazed, tilled or traversed.

But now for the hard part. Let there be no illusions; not very many people will be in favour of the idea. This puts the development assistance industry in a bind; they can only suggest, not impose.

Perhaps 75% or more of Africans are still rural inhabitants, and peasants are properly suspicious of anything that will demolish an ancient heritage, with its comforting tradition of extended family responsibilities. The chiefs along with the government (more strictly the president; it is his own government) are also fundamentally opposed, the former overtly because it removes their main social function, the latter covertly because it removes political clout. It may seem paranoid to accuse administrations of wishing to keep their citizens poor and hence subservient, but a top

A thought at sundown

(black) civil servant in Zimbabwe gloomily agreed with me when I suggested that this seemed to be the only reason for that government's refusal to give full title to communal peasants resettled on former freehold land.

Yet probably many younger men can be counted on to support change; their responses to my inquiries suggest that they would cheerfully dispense with polygamy, which is something wealthier, older men have an advantage in. Girls too, and not just educated ones, are almost unanimous in their rejection of a system that makes them mere property. In addition, many ordinary people in the cities approve such changes, even if they are not already members of the propertied classes, for they have *chefs* as rôle models all about them, while their links with their rural culture are weakening.

On the other hand, though, it is very convenient to have a wife, or better, girlfriends, looking after you in the city and another spouse producing children and tending livestock in the communal countryside. Numerous *chefs* agree, it seems.

Beyond this an even greater problem looms.

•   •   •   •

On the UNDP's Web site for Human Development there is a country by country table of military expenditures, ranked in the order that they fall in the UNDP's Human Development Index (*www.undp.hrdo/military.htm*). The last forty-four of these (Myanmar down to Sierra Leone) come under the heading Low Human Development. Thirty-four of these are in Africa and I have worked or travelled in twenty-three of them.

Logic suggests that military expenditure as a proportion of that spent on education and health would fall in the poorer countries. Experience suggests otherwise, and this is borne out by the numbers. The figures are out of date—they apply to 1990–91—but they show that at the end of the cold war, developed countries were spending a figure equivalent to about 38% of their combined health and education budgets on the military. For developing countries overall the figure was 63%. However, for the least developed countries, which, as we have seen, are mainly in Africa, the figure was 73%.

Even this is probably an understatement. It came out during a parliamentary enquiry in Uganda in 1999 that part of the defence ministry budget is distributed amongst other ministries, from whence it is forwarded to the army. A senior civil servant, giving evidence, said that this is done because the government does not want trouble from aid donors who insist on limits to military spending.[154] Even more egregious was the surfacing of evidence in Zimbabwe in October 1999 that it had spent about ten times as much on Robert Mugabe's military adventure in the Democratic Republic of the Congo in the first half of that year as it had told the IMF it was.[155] On the basis of the smaller figure the IMF had extended a loan of $193 million to Zimbabwe; it turned out that the amount actually disbursed on the war during those six months was of the same order. Further loans to Mugabe's government were suspended.

In passing, there seems to be no case in Africa where war has been formally declared following a debate in a country's parliament, even if the latter is packed with the president's henchmen. Even Richard Nixon had to defend his move into Vietnam in Congress with an overblown story about a destroyer attacked in the Gulf of Tonkin. Such retrospective justification is not mandatory in Africa.

African armies, it is probably unnecessary to say, are seldom disciplined bodies of men with proud regimental histories and a discernible *esprit de corps*. No, such forces are primarily agents of the country's leader, usually tribally affiliated and owing their perks (good pay and housing, duty-free liquor, plentiful rations) to him. In return they do his bidding, going to war with whomever he tells them to attack, scouring their own nation if necessary. Of course, if the flow of goodies falters then so does loyalty. Between 1970 and 1995 there were over fifty military coups in sub-Saharan Africa, with perhaps another ten up to the end of the century. Interestingly, almost all of these were started by junior officers or lower ranks, dissatisfied over their prospects or pay, or by the rise of influence of a rival tribe in the army.[156] The generals, by and

A thought at sundown

large, are not too keen to rock the trough they feed from (although they usually accede graciously enough when their juniors select them to be the next head of state).

For this reason most military upheavals in Africa fail. The total of successful coups understates the total number of disruptions that Africa's big armies cause to civil society by perhaps two-thirds.

There lies the big difficulty with the nice idea that through extending land ownership to the people of Africa they will get richer. The sanction of violence that an undisciplined army creates—not necessarily a large army, even—makes a mockery of attempts to achieve the rule of law needed to a create a stable, bourgeois, property-owning society.

There are therefore five main stumbling blocks to prevent Africa from moving forward to prosperity through *embourgeoisement*. They are the president, his army, the chiefs, the peasants and the donors.

• • • •

The donors? Well, they do not appear to be part of the solution so far; if they were then three hundred billion dollars would have something to show for itself. Compared with other parts of the world, in Africa their support for land tenure reform has been equivocal, presumably because governments are not very interested in such a fundamental change. This has meant for donors that the chance of failure during implementation will be high and, at the least, careers could be seriously damaged.

Yet the aid agencies' commendable desires regarding governance and civil society can only be met if the threat of anarchy is removed and *embourgeoisement* commences. This threat comes from two sources, the first being the proliferation of weapons in African societies. The continent is awash in (mainly) eastern-bloc assault rifles, the ubiquitous AK47 leading the list. In country after country where I have worked, weapons have been on offer at amazingly low prices: $50 for an AK in Lichinga (Mozambique), $20 for a Taurus revolver in Uganda, $15 for a

All poor together

Tokarev pistol in up-country Mali.

These are just the surplus to the holdings of the various breakaway fractions that live a shadowy half-life in most countries: the Tuaregs at the desert-savannah interface along a band stretching from Mali to Chad, the coast/interior confrontations in Sierra Leone and Liberia, the tribal factions in the Congo (Brazzaville), the infamous Hutu/Tutsi divide in the Great Lakes, the endless Arab-African war in the Sudan, the Somalis versus everybody else.

To name a few. Yet to rid Africa of its illegal weapons should be the easier of the two destabilizing threats. It would, after all, have the support of the 'legitimate' authorities in the respective countries and could be the subject of various United-Nations-brokered agreements, universal registration, a multinational inspectorate, perhaps a coastal blockade and so on. The second problem is much more difficult: it is to neutralize the threat from the weapons held by the legal holders—the army.

While the reduction of the threat of civil violence from illegal weaponry will help, it will not be possible to remove the army: its function is to maintain the big man in power. For it to be dissolved the main problem is not the resistance of the army to the idea, it is the resistance of the president. But that is not the objective anyway, the objective is to prevent the army from undermining the rule of law; from mutiny and rapine. This is achieved by keeping it in the barracks, which is achieved by paying it well.

How to find the money? Let us stand the development assistance rationale on its head. One of its major concerns is that aid money is fungible. Like water it does not compress easily, so the addition of money at one end of a government, perhaps in education, will (in theory at least; it tends to leak out elsewhere) displace the equivalent amount for use elsewhere in that government. In the army, say. It is not an academic concern; there is a fair body of knowledge which confirms that this happens. The case of Zimbabwe's IMF loan above may stand as an example. A recent paper[157] on a sample of 38 developing countries found that most aid appears to be fungible; countries receiving concessionary

A thought at sundown

loans for development in agriculture, energy and education reduced their own input to these sectors and used the money so liberated elsewhere. The authors did note that—

> Data from our sample countries do not support the hypothesis that foreign aid is being diverted for military purposes.

But this is not at all the same thing as saying that money freed up by aid funds is not fungible in relation to military expenditure.

But suppose we go to the other end of the pipeline, to the military. If they are seen, not as soldiers but as *peacemakers*, then perhaps donors have the argument needed to take direct responsibility for the payroll there, thus displacing local money to the other, more productive outlets of government. (Joking? I am not joking). This could be done with the full panoply of bureaucratic controls on disbursements that usually hamper aid projects, so this support of the soldiers would be, if not a transparent process, a fairly uncorrupt one. At the same time a whole host of problems falls away. Not only the fungibility problem but the keeping-them-in-the-barracks problem, the controlling-the-size-of-the-army problem, even to an extent the buying-fancy-and-unnecessary-new-hardware problem.

More, the soldiers could be the first beneficiaries of land tenure reform, following an ancient tradition of rewarding loyal veterans with grants of freehold land. Many of the soldiers who so benefit may be likely to conform to Joe Cohen's prediction, but at least their expenditures will benefit the local economy, even if it is the brewery.

A funny thing for donors to do? If their objective is to create an environment in which 'good governance' and a 'civil society' exists, then 60-odd successful military coups and perhaps 100-plus failed ones in 35 years, not counting the many African border conflicts as such, suggests that this is the only sure way.

They are already deep in this business anyway. The United States, Britain and France have a score of military assistance and training programmes in Africa. For example, U.S. military assistance to Kenya in the 1980s included grants from the Military

All poor together

Assistance Program (MAP) loans from the Foreign Military Financing (FMF) and International Military Education and Training (IMET) allocations. IMET is a program by which foreign military personnel attend U.S. military schools. IMET funds averaged more than $1 million a year from 1986 to 1990. Funding for MAP and FMF has exceeded $250 million since 1975.[158]

In 1998 France provided $30 million in aid to African states to strengthen their peacekeeping capacity under its RECAMP program (Reinforcement of African Peace-keeping Capabilities). In the same year the 'Guidimakha' exercise brought together more than 3,500 soldiers from eight African countries. In terms of pre-positioned forces, contingents from the French navy, air force and army are, or were, based in the Ivory Coast, Senegal, Gabon, Djibouti, the Central African Republic and Chad. The extent of French military involvement can be gauged by the fact that at one point in 1995 it had around 7,000 troops in Africa.[159]

The British have been more low-key. They sponsored the development of the Zimbabwe Staff College into what is claimed to be the world's first 'regional centre for excellence'. Another such is being developed in Ghana. A multinational peace-keeping training exercise called Blue Hungwe was held in Zimbabwe in April 1997, involving around 1,400 military personnel from ten SADC countries.[160]

Even the Danes (the Danes!) in conjunction with the Zimbabwe army (the Zimbabwe army!) have commissioned a 'Centre for Regional Conflict Resolution' in Harare. Its effect on that country's military adventure into the Democratic Republic of the Congo in support of Laurent Kabila has not, as yet, been noticeable.

The funding of Africa's armies could be best administered through the UN, as African states have already made a major contribution to UN peace-keeping. It was calculated in 1995 that 22 African countries had participated in 21 UN peace-keeping operations since 1960. This amounted to 96 individual contributions from African countries to these operations, with Ghana, Egypt, Nigeria, Senegal and Tunisia having a total of 51 contributions (some 53% of the total).[161]

Peace-keeping jobs are sought after by African states because

A thought at sundown

the UN usually pays the country significantly more for the soldier's services than they actually cost. However, the fragmented bilateral efforts by the developed world to have a say in the African military by providing training should be consolidated into a single fund and the job handed over to the development assistance people. They would disburse money directly, the intention being to turn the soldiery into fat and happy peace lovers. Donors would also guarantee the borders and fund boundary commissions to ensure that no disputes disturb the even tenor of military life. This change of emphasis, properly explained, would be welcomed by the taxpayers and voters of the donor countries.

Perhaps just a couple of rules: outside the barracks and training grounds, no uniforms except for generals and no weapons either, except on ceremonial occasions. No ammunition to be issued unless the president gets agreement from the donors. If he breaks these rules, then he will suddenly have an unpaid army on his hands.

Now fungibility works in the donor's favour; money is freed up within government for such worthy sectors as education and health, for governance, for land reform even. The problems of both the donors and the army have been solved. That leaves the president and his government, the peasants and the chiefs.

In fact the president can be by-passed. He is still impressively buttressed by portly generals emblazoned with gold braid and service ribbons, but his army is toothless. It only requires that he remains passive while the revolution carries on below, and that can be achieved by ensuring that he, his government and his army are amongst the earliest beneficiaries of land reform.

The chiefs. They are the most difficult obstacle of all, because they are the guardians of the culture that has caused the problem. They may not understand what is going on, but they will know that it means the end of their authority. They are already sensitive about this because governments never give traditional leaders the attention they believe they merit. Politicians are polite but distant to them, fearful that they will bring up complaints about the inadequacy of their stipends yet again.

The answer seems to be to institutionalize what is already

All poor together

happening in many parts of Africa. Population pressure has meant that communal land has started to have a value, and chiefs, elders and headmen, as well as the peasants themselves, have taken to selling it to incomers.[162] The new regime will take away that right from the customary leaders, and it must be replaced with something else. That something could be related to each land transaction that occurs in what used to be his domain, perhaps a local duty on each transfer, say. Or a bottle of gin. To put it bluntly, the chiefs will have to be suborned; there is no avoiding their resistance otherwise.

Finally, the peasants. Here is where the attitude of the chiefs is the key. If the rural dwellers see that the chiefs are themselves selling land with official approval, and sense that the government is no longer hostage to its men of violence, then they will begin to think in turn about the need to secure the rights to the land they occupy.

There would be a carrot to this stick. The new dispensation will encourage them to plan beyond the next rains. A well, a grain mill, a small tractor, even. But to make it possible for them to do something about it, they will need a loan, for which they will require security. Tenure—freehold or long-term leasehold—would provide this. If both tenure and loans are conveniently available on request, if there is a land titling agency in the village, and a bank as well, then it would be easy to blunt the eldest wife's nagging about needing a new plough by going to find what it is all about.

The success of this programme can be judged quite simply by the appearance of a novel phenomenon in rural Africa. Estate agents.

• • • •

This has not, I hope, been too weary a progression. The evolution of that brash innocent, who hoped to impress Stamico (*Stamico!*) by putting his finger in concentrated nitric acid, to the despairing cynic of today has had its jolly moments. But by now the nightmare creature forming behind this tale of who found what and who did what should be bursting through the painted scenery.

The lesson is this. We, who have struggled out of serfdom into

A thought at sundown

a situation where food, drink and shelter are no longer our main concerns are, like it or not, creatures of the Age of Reason. Newton, Leibnitz, Voltaire and, belatedly, Adam Smith, have completed the victory of rationality. The emergence of America, Canada, Australia and other rude, practical societies, democratically ruled by their rude, practical inhabitants, drove out the old structured, irrational, custom-following, religion-haunted ethos of humankind. Perhaps—one can only hope—for ever. Man became mortal man. Man became economic man.

Not, however, in Africa. To get a flavour of life today in Kinshasha, in Lagos, in Nairobi, read Pepys' unexpurgated diaries. The financial and sexual corruption, the fawning dependency upon a big man, the untrustworthiness of anybody outside one's immediate circle, one's family—and not even then—are replicated every day in the societies of those cities.

My arrogance in saying what follows pains me, but I believe it true. African man is not economic man, he is still cultural man, embedded in an aggressively structured, profoundly conservative society that has evolved to enable him to survive in a vast, barren continent with only iron-age technology to help. Consciously or unconsciously, the method we have adopted to change his thinking is called development assistance, and it rests on the assumption that what pulls together all the strands of such work—education, health, infrastructure, agriculture and so on—is the economics of a nation state. Macroeconomics, it is called. Yet this science is not merely dismal but meaningless to somebody whose principal aim in life is to acquire enough livestock to buy another, younger, more fecund wife, or to find the money, somehow, to meet his obligation to put a younger brother through school, or to get a safe job in government through his aunt's husband's great-uncle, the minister.

The very idea of trying to alter another race's culture is abominable to us, reeking of the foul acts towards other peoples that have smirched our history. It is almost unthinkable.

But it has to be thought. As Galbraith noted, we had capital and technology and poor countries did not. Therefore we concluded poverty could be alleviated by transferring these from us to them.

All poor together

It was not, and is not, so easy. Our real wealth is in the ideas—never fixed—about society which make the rapid creation of wealth and technology possible. Independent, individualistic, meritocratic, capitalist and fixed property-owning. In Africa almost no attempt has been made to get the continent to accept them.

There are now about twenty-five times as many people in Zimbabwe as there were in 1900; by 2020, AIDS permitting, there will be forty times as many. There are two major tribal groupings. Rwanda has shown us what the cultural outcome is when that unstoppable progression hurtles into the immovable limits of a country's resources.

* * * *

The briefing was coming to an end.

'Questions?'

Somebody whose reading must have extended beyond *The Rhodesia Herald* asked about strikes by workers in the United Kingdom. Did he agree that Callaghan's government was going to fall?

'Now don't you people go thinking that because the Conservatives might get into power that the Brits will change their tack. I hear that this Margaret Thatcher is a real tough bitch.'

The lecturer rolled up his maps. Yorky, the chief instructor, reappeared.

'Fall in!' We assembled untidily, languid in the evening warmth.

'Look smart there! Come on you lot. Christ! Fred Karno's army ... Right. Orders. Embus at eighteen hundred hours. That's only ten minutes so get your kit together right after you fall out and double down to the road. There'll be scoff at Fort Vic and you will draw more ammo there. ETA midnight, so fill your water bottles now ...'

Stumbling down through the bush under the weight of our kit towards the short row of snub-nosed Bedfords, a mate of mine called Tony muttered darkly to me about how the politicians and

A thought at sundown

the FGs were going to screw this country up.

'You heard what that guy said, about Tanzania and so on. Let them take over and they'll do that to us as well.'

Tony was killed a few months later, when the armoured vehicle he was in went over a land mine. It comes to me while writing this that there was some story about him not being strapped in. This was a serious offence, as it often made the difference between life and death. Because of that, it was said, his widow might not get a pension. I am sure she did in the end, but I doubt if it is worth anything now.

'You've been about a bit, John. What do you think?'

What did I think? The dry season was approaching its end and the *mfuti* trees on top of the kopje opposite already had their salmon-pink canopies of new leaves, now glowing in the setting sun. A slight smell of woodsmoke hung in the air and fine fragments of burnt grass drifted down. It was difficult to be pessimistic. Spring was here, soon the rains would come.

'I think we'll be all right,' I said.

REFERENCES

1. Nhachi, C. F. B. 'Cases of Poisoning in Zimbabwe: A Review', *Zimbabwe Science News*, Vol. 30, No. 4, October–December 1996, pp. 101–4.
2. Psacharopoulos, George *Journal of Human Resources* (Fall of 1985), pp. 583–604.
3. Vice President for Africa (Edward V. K. Jaycox): Statement to the Board of the World Bank, November 19th, 1990
4. World Development Report, 1998/99, World Bank, Washington, 1999.
5. *Africa Recovery*, April 1992, p. 13, UNDP, New York
6. 'The IMF and the Poor', Pamphlet Series No. 52 (Washington: IMF, 1998).
7. Zimbabwe Statistical Yearbook 1997, Central Statistical Office, Harare, July 1998.
8. UNESCO, World Education Report, 1991.
9. Carrington, W. and Detragiache, E. *How Extensive is the Brain Drain?* Finance and Develop-ment, IMF, June 1999.
10. *World Development Report on Poverty,* World Bank, 1990.
11. Nyerere, J. Article in *The Guardian*, 10 June 1997.
12. Zimbabwe Statistical Yearbook 1997, Central Statistical Office, Harare, July 1998.
13. World Development Report 1998/99, *Knowledge for Development*, The World Bank, 1999
14. Castro-Leal, F., Dayton, J., Demery, L., Mehra ,K. *Public Social Spending in Africa: Do the Poor Benefit?* The World Bank Research Observer, Vol. 14, No. 1 (February 1999), pp. 49–72.
15. Ibid.
16. Ghanem, H. and Walton, M. 'Workers Need Open Markets and Active Governments', *Finance and Development,* September 1955.
17. Auty, R. M. *Zambia's Mismanaged Mineral Independence*, Resources Forum, Vol. 17, No. 3, pp. 170–183.
18. *World Development Report, Investing in Health*, World Bank, Washington, 1993.
19. Castro-Leal, F., Dayton, J., Demery, L., Mehra, K. *Public Social Spending in Africa: Do the Poor Benefit?* The World Bank Research Observer, Vol. 14, No. 1 (February 1999).
20. *World Development Report, Investing in Health*, World Bank, Washington, 1993.
21. McLaughlin, Peter *The Ragtime Army,* Books of Zimbabwe Publishing Co., Bula-wayo, 1980.
22. *World Development Report,* World Development Indicators, Table 7, World Bank, Washington, 1999.
23. White, Howard *Is development aid harmful to development? The Courier*, January–February, 1993.

References

24. *The Aids Epidemic and its Demographic Consequences*, United Nations/WHO, New York, 1991.
25. Timaeus I., 'Adult Mortality', in *Demogaphic Change in sub-Saharan Africa*, National Academic Press, Washington D.C. 1993.
26. Bulatao, R. *Projecting the Demographic Impact of the HIV Epidemic Using Standard Parameters,* United Nation/WHO, *The Aids Epidemic and its Demographic Consequences,* New York, 1991.
27. Anderson, R. M. 'The Impact of the Spread of HIV on Population Growth and Age Structure in Developing Countries', in *The Global Impact of Aids*, New York, Alan R. Liss, Inc. 1988.
28. Ainsworth, M. and Mead, O. 'Aids and African Development', *The World Bank Research Observer,* Vol. 9, No. 2, 1994, pp. 203–240.
29. Ibid.
30. *World Development Report, Investing in Health*, World Bank, Washington, 1993.
31. Harvey, Martin, Southampton Insurance Company of Zimbabwe, 1999 Southern African Economic Summit, Durban South African, July 1999.
32. Whiteside, Alan, Director, Health Economics and HIV/AIDS Research, University of Natal, South Africa, Southern African Economic Summit, Durban South African, July 1999.
33. Derbyshire, S. 'The Truth About the AIDS Panic in Africa', *Living Marxism,* issue 79, May 1995.
34. 'The incalculable cost of AIDS', *The Economist,* March 12th, 1998, p. 68.
35. Madavo, C. and Sarbib, J.-L. *Memorandum to World Bank Staff and Supporters,* 2 June 1999.
36. *WHO Weekly Epidemiological Records No. 69,* 1 July 1994.
37. Cleaver, K. M. and Schreiber, G. A. *Reversing the Spiral. The population, agriculture and environment nexus in sub-Saharan Africa*, The World Bank, 1994.
38. Cleland, J. and Hobcraft, J. *Reproductive change in developing countries: Insights from the World Fertility Study*, Oxford University Press, London, 1985, pp. 273–95.
39. *A Continent in Transition: sub-Saharan Africa in the Mid-1990s*, The World Bank African Region, Washington, November 1995.
40. Dove, Aland *PICO News,* 7 June 1999.
41. Baines, T. *The Gold Regions of South Eastern Africa, With Portraits, Maps, Illustrations*, FRGS, London, 1877.
42. Goossens, P. J. *Or de l'Antiquit a nos jours*, Societe de l'Industrie Minrale, Paris, 1998.
43. *The Discovery of Gold on the Witwatersrand,* Report of the Committee of Enquiry appointed by the Commission for the Preservation of Natural and Historical Monuments, Relics and Antiques, Pretoria, February 1941.

44 Ibid.
45 *Wonderful South Africa*, Associated Newspaper Ltd, Johannesburg, 1936?
46 Williams, Alpheus F. *Some Dreams Come True*, Howard Timmins, Capetown, 1948.
47 Ibid.
48 Cartwright, A. P. *The Gold Miners*, Purnell and Sons (SA) Pty Ltd, Johannesburg, 1962.
49 Letcher, Owen *The Gold Mines of Southern Africa*, Published by the author, June 1936.
50 Pers. comm. Barplats Management, 1990.
51 Hammond, J. H. *The Autobiography of John Hays Hammond*, Vols. I and II, Farrar and Tinehart Inc., New York, 1935.
52 Bowen, D. J. *Gold Mines of Rhodesia, 1890–1980*, Thomson Newspapers (Zimbabwe) Ltd, Harare, 1980.
53 'Busang Excitement Mounts', *Mining Journal*, 26 January 1996.
54 Mercury Gold and General Fund Newsletter, 7 June 1996.
55 *Forensic Investigative Associates Incorporated (FIA) Report*, DeloitteTouche Inc., Trustee in Bankruptcy of Bre-X Minerals Ltd., Toronto, Canada, 1998.
56 The story of what probably happened in the final days of the Bre-X fraud has been reconstructed from the files of the *Northern Miner*, the *Mining Journal* and the Forensic Investigative Associates Inc. report referred to above.
57 Lawrence, M. J. *Project Evaluation Due Diligence—lessons from the Busang Saga*, World Gold '97 Conference, Singapore, 1997, pp. 249–264.
58 Crowson, Philip *Mineral Handbook 1996–1997*, Mining Journal Books, 1997.
59 Folie, Dr G. Michael *Gold Mining in Today's Price Environment*, Acacia Resources Ltd, 22nd Annual *Financial Times* World Gold Conference, June 1999.
60 Shute, Nevil 'Norway', *Slide Rule*, (Chapter 7 of his autobiography) 1954.
61 *Bank's World*, 10 October 1997.
62 Boone, P. *The impact of foreign aid on savings and growth* and *Politics and the effectiveness of foreign aid*, The Centre for Economic Performance, London School of Economics, Working papers 1265 and 1267, 1994.
63 *See*, for instance, Krueger, A. O., Michalopoulos, C. and Ruttan, V. *Aid and Development*, Baltimore and London, Johns Hopkins University Press, 1989.
64 *Assessing Aid. What Works, What Doesn't and Why*, The World Bank, Washington D.C., 1998.

References

65 Ibid.
66 *World Development Indicators 1997*, The World Bank, Washington D.C., 1997.
67 Schumpeter, J. A. *The Theory of Economic Development*, Harvard University Press, 1934.
68 This chapter draws upon Waalbroek, J. 'Half a Century of Development Economics: a Review Based on the Handbook of Development Economics', *The World Bank Economic Review*, Vol. 12, No. 2, pp. 323–352, 1998.
69–70 Ibid.
71 *See*, for instance: Rostow, W. 'The take-off into self-sustaining growth', *Economic Journal*, Vol. 66, March 1965, pp. 24–68.
72 Eberstadt, Nicholas 'Foreign Aid's Industrialized Poverty, *Wall Street Journal*, Wednesday, 8 November 1989.
73 Galbraith, J. K. *The Nature of Mass Poverty*, Harvard University Press, 1979.
74 Marsden, K. *Africa's Entrepreneurs*, Discussion Paper, International Finance Corporation, Washington D.C., 1990.
75 'Africa: a Continent at Stake', *Financial Times*, September 1, 1993.
76 Santa Cruz, Hernan *Réforme agraire et dévelopement rural*, Forum du développement, May 1979, p. 1.
77 Waalbroek, J. 'Half a Century of Development Economics: a Review Based on the Handbook of Development Economics', *The World Bank Economic Review*, Vol. 12, No. 2, 1998, pp. 323–352.
78 Burnside, C. and Dollar, D. 'Aid Spurs Growth—in a Sound Policy Environment', *Finance and Development*, December 1997.
79 'Sick patients, warring doctors', *The Economist*, London, September 18, 1999.
80 Aziz, J. and Wescott, R. F. *Policy Complementaries and the Washington Consensus*, IMF Working paper O38, 1997.
81 Mbanefo, Uche *The Origins of the Need for Economic Adjustment in Africa*, Seminar on Zambia Economic Adjustment and Reform Programme, Livingstone, Zambia, January 1987.
82 *The Herald*, Harare, Saturday, 25 September 1999.
83 *The Economist*, 23 September 1999.
84 Burnside, C. and Dollar, D. *Aid, the Incentive Regime and Poverty Reduction*, World Bank, 1998.
85 Waalbroek, J. 'Half a Century of Development Economics: a Review Based on the Handbook of Development Economics', *The World Bank Economic Review*, Vol. 12, No. 2, 1998, pp. 323–352.
86 'Diamonds are for … smuggling', *Mining Journal*, London, 2 August 1996, Focus and Comment, p. 93.
87 Mackay, R. A. 'A year of digging on the Lupa Goldfield', *The Mining Magazine*, February 1948, pp. 73–76.

88 McGill, S. C. *Legislative and economic policies for the promotion and regulation of small scale mining in an age of international mining capital*, UNDTCD, 1984.
89 *Decret No. 91–277, Article 36, Republique du Mali.*
90 Ibid., *Article 52.*
91 See, for instance, *Report on the workshop on ecologically sustainable gold mining and processing,* UNIDO Project No. XP/INT/95/043, Jakarta, November 1995.
92 Campbell, S. D. G. and Pittfield, P. E. *Structural Controls of Gold Mineralization in the Zimbabwe Craton,* Bulletin No. 101 of the Zimbabwe Geological Survey, 1994.
93 *The Economist,* 4 September 1999.
94 Mackay, R. A. 'A year of digging on the Lupa Goldfield', *The Mining Magazine,* February 1948, pp. 73–76.
95 Dahl, Roald *Going Solo,* Jonathan Cape Ltd, London, 1986.
96 Royle, A. G. 'The Lupa Goldfield, Tanzania', *Leeds University Mining Assoc. Journal,* 1995.
97 Ibid.
98 Tan Discovery Minerals Consulting Co., Ltd, *Baseline survey and preparation of development strategy for small-scale and artisanal mining program,* Dar-es-Salaam, June 1996.
99 Crisp, K. A. 'Currency volatility—its impact on both supply and demand for gold', *Financial Times'* World Gold Conference, London, 14–15 June 1999.
100 Warren, George F. 'Some Statistics on the Gold Situation', *The American Economic Review,* Vol. XXIV, No. 1 (Supplement of March 1934).
101 *Gold '98,* Goldfields Mineral Surveys, London 1998.
102 Milling-Stanley, George 'Marketing and Demand Issues: Setting the Scene', *Financial Times'* World Gold Conference, June 1999.
103 Churchill, Lord Randolph *Men, Mines and Animals in South Africa,* Sampson Low, Martson and Co., London, 1893.
104 Garlake, Peter S. *Great Zimbabwe,* Thames and Hudson, London, 1973.
105 Iliffe, J. *The Africans,* Cambridge University Press, 1995.
106 Burke, R. E. (ed.) *The Journals of Carl Mauch,* Salisbury, 1969.
107 Gayre of Gayre, R. *The Origin of the Zimbabwean Civilization,* Galaxie Press, Salisbury, 1972
108 Huffman, T. N. and Vogel, J. C. 'The chronology of Great Zimbabwe', *South African Archaeological Bulletin,* **46** (1991), 61–70.
109 Summers, R. *Ancient Mining in Rhodesia and Adjacent Areas,* Nat. Mus. Rhodesia Mem. 3, Salisbury, Trustees of the National Museum of Rhodesia, 1969.

References

110 Swan, L. 'Early gold mining on the Zimbabwe plateau. Changing patterns of gold production in the first and second millennium AD', *Studies in African Archaeology* **9**, Societas Archaeologica Upsaliensis, 1994.
111 Garlake, Peter S. *Great Zimbabwe*, Thames and Hudson, London, 1973.
112 Herbert, E. W. 'Metals and power at Great Zimbabwe', in *Aspects of African Archaeology, Papers from the 10th Congress of the Pan African Association for Prehistory and Related Studies*, by Pwiti, G. and Soper, R. (eds) Harare: University of Zimbabwe Publications, 1996.
113 Garlake, Peter S. *Great Zimbabwe*, Thames and Hudson, London, 1973.
114 Beach, D. N. *Archaeology and History in Nyanga, Zimbabwe*, Seminar paper presented to the History Department, University of Zimbabwe, 1995.
115 Chirawu, S. 'Ancient Terrace Farming in North-Eastern Zimbabwe', *Zimbabwe Science News*, Vol. 31, No. 3, July/September 1997.
116 *Mining Journal*, London, 8 May 1998, p. 365.
117 'Silicosis link to lung cancer', *Mining Journal*, London, December 18/25, 1992.
118 Ibid.
119 *British Asbestos Newsletter*, Issue 19, Spring 1995.
120 WHO *Environmental Health Criteria #53*, Geneva, 1986 (p. 133).
121 Montague-Murray, H. M. *Report of the Department Committee on Compensation for Industrial Disease*, London, 1907.
122 Cooke, W. E. 'Pulmonary asbestosis', *Brit. Med. J.* **11**, 1927, p. 1024.
123 Lanza, A. J., et al. 'Effects of the inhalation of asbestos dust on the lungs of asbestos workers', *Public Health Reports*, **50**, 1935, pp. 1–12.
124 McDonald, J. C. and McDonald, A. D. 'Chrysotile, Tremolite and Carcinogenicity', *Ann. Occup. Hyg.*, **41**, No. 6, 1997, pp. 669–705.
125 EPA 1984, *Asbestiform Fibers: Nonoccupational Health Risks*, Prepared by a committee of the National Research Council with support under Contract No. 68–01-4655 with the Environmental Protection Agency, Washington, 1984.
126 Selikoff, I. J., Churg, J. and Hammond, E. C. 'Asbestos exposure and neoplasia', *J. Am. Med. Assoc.* **188**, pp. 22–26.
127 Wagner, J. C., Sleggs, C. A. and Marchand, P. 1960, 'Diffuse pleural mesothelioma and asbestos exposure in the North Western Cape Province', *Br. J. Ind. Med.* **17**, 260–271.
128 Morello, Carol *USA Today*, 10 February 1999.
129 Langer, M. and Nolan, R. P. *Fiber Type and Mesothelioma Risk*, Symposium on Health Effects of Exposure to Asbestos in Buildings, Harvard University, December 14–16, 1988.

130 Jeffreys, D. B. and Vale, J. A. 'Malignant mesothelioma and gas mask assemblers', *British Medical Journal,* 1978 (26 August); 2: p. 607.
131 Mossop, R. 'Asbestos Hazards in Zimbabwe', *Cent.Afr.J.Med.* **29**, No. 5, May 1983, pp. 117–8.
132 McDonald, J. C., et al. 'The 1891–1920 birth cohort of Quebec chrysotile miners and millers: mortality 1976–1988', *Br. J. Ind. Med.* **50**, 1993, pp. 1073–1081.
133 Smith, A. and Wright, C. 'Chrysotile Asbestos is the Main Cause of Pleural Mesothelioma', *American Journal of Industrial Medicine,* **30**, pp. 252–266, 1996
134 Davis, J. M. G. *Evidence for Variations in the Pathogenic Effects of the Different Forms of Commercially Used Asbestos, A Review of the Literature,* Edinburgh, Scotland: Institute of Occupational Medicine, Pathology Branch, 1980 (December); Report No. TM/80/4, UDC No. 616 553.676: 27.
135 McDonald, A. D., et al. 'Mesothelioma in Quebec chrysotile miners and millers; epidemiology and aetiology', *Ann. Occup. Hyg.* **41**, 1997, pp. 707–719.
136 McDonald J. C. and McDonald, A. D. 'Chrysotile, tremolite and carcinogenicity', *Ann. Occup. Hyg.,* Vol. 41, No., 6, pp. 699–705, 1997.
137 Amandus, H. E., et al. 'Mortality of vermiculite miners exposed to tremolite', *Ann. Occup. Hyg.* **32**, 1988, pp. 459–465.
138 Armstrong, B. G., et al. 'Radiological changes in vermiculite miners exposed to tremolite', *Ann. occup. Hyg.* **32**, 1988, pp. 469–473.
139 McConnochie, et al. 'Mesothelioma in Cyprus: the Role of Tremolite', *Thorax,* 1987; 42: 342–347.
140 Piloatto, G., et al. 'An update of cancer mortality amongst chrysotile miners in Balangero, Northern Italy', *Br. J. Ind. Med.* **47**, 1980, pp. 810–814.
141 Pers. Comm. Brian Gibson, Johannesburg, 20 October 1999.
142 Wagner, Dr J. C. Acceptance speech on receiving the Charles S. Mott prize, General Motors Cancer Research Foundation, 1986.
143 Letter to the Zimbabwe Minister of Industry and Commerce, August 1999.
144 *IDA in Action 1993–1996: The Pursuit of Sustained Poverty Reduction,* The World Bank, 1997.
145 Chandiwana, S. and Shiff, C. 'Science-based Economic Development: the Eureka Factor', *Zimbabwe Science News,* Vol. 33, No. 1, January–March 1999, pp. 5–12.
146 'Entering the 21st Century,' *The World Bank World Development Report 1999/2000,* Box 6.4, The World Bank, Washington D.C., p. 131.

References

147 Deininger, K and Binswanger, H. 'The Evolution of the World Bank's Land Policy: Principles, Experience and Future Challenges', *The World Bank Research Observer,* Vol. 14, No. 2 (August 1999), pp. 247–276.

148 Binswanger, H. and Pingali, P. L. *Resource endowments, farming systems and technological priorities for sub-Saharan Africa,* Discussion paper ARU 60, Agricultural and Rural Development Department, World Bank, 1987.

149 Cleaver, K. M. and Schreiber, G. A. *Reversing the spiral. The population, agricultural and environmental nexus in sub-Saharan Africa,* The World Bank, 1994.

150 Migot-Adholla, S. E., Place, F. and Oluoch-Kosura, W. 'Security of tenure and land productivity in Kenya', in *Searching for Land Tenure Security in Africa,* The World Bank, 1993.

151 Summers, L. H. and Thomas, V. 'Recent lessons of development', *The World Bank Research Observer,* Vol. 8, No. 2 (July 1993), pp. 241–54.

152 *Reversing the spiral. The population, agriculture and environment nexus in sub-Saharan Africa,* Box 11.1, World Bank, 1994.

153 'The African Condition', *The Chartered Secretary,* p. 23, Harare, 4th Quarter 1999.

154 'Creative accounting in Africa: hidden skills', *The Economist,* London, 9 October 1999.

155 *The Financial Times,* London, 4 October, 1999.

156 Banks, A. S. *Cross-National Time Series Data Archive,* Center for Social Analysis, SUNY, 1994.

157 Feyzioglu, T., 'Swaroop, V. and Zhu, M. 'A panel data analysis of the fungibility of foreign aid', *World Bank Economic Review* 12 (1), 1998.

158 Prinslow, Lieutenant-Colonel Karl E. 'Building Military Relations in Africa', *US Army Military Review,* Vol. LXXVII, May–June 1997, No. 3.

159 Ginifer, Jeremy *Emergent African Peace-keeping: Self-help and External Assistance,* UN Programme at the Norwegian Institute of International Affairs (NUPI), Oslo, 1999.

160 Ibid.

161 Ibid.

162 See, for instance, Chapter 11 in *Searching for Land Tenure Security in Africa,* Bruce, J. W. and Migot-Adholla, S. E., The World Bank, 1993.